INSECTS AND FLOWERS

INSECTS AND FLOWERS
A Biological Partnership

JOHN BRACKENBURY

BLANDFORD

A BLANDFORD BOOK

First published in the UK 1995 by Blandford
An imprint of Cassell
Wellington House
125 Strand, London WC2R 0BB

Distributed in the United States by Sterling Publishing Co., Inc.,
387 Park Avenue South, New York, NY 10016-8810

Distributed in Australia by Capricorn Link (Australia) Pty Ltd
2/13 Carrington Road, Castle Hill, NSW 2154

A Cataloguing-in-Publication Data entry for this title is
available from the British Library

ISBN 0-7137-2491-9

Typeset by Litho Link Ltd, Welshpool, Powys, Wales
Printed and bound in Hong Kong by Dah Hua Printing Press Co.

CONTENTS

INTRODUCTION

ONE HOT SUNNY day in June I was standing beneath a grey-leaved willow tree, and became aware of large blobs of rain landing on my head and clothing. This was quite puzzling because there wasn't a cloud in the sky. It took me quite some time to work out where the rain was coming from. Then I noticed the flecks of cuckoo-spit, hanging pendulously from the branches, the work of the nymph of the froghopper *Aphrophora*. Every now and then, a large droplet would swell up inside the frothy mixture and suddenly detach itself. Taken across the tree as a whole this amounted to a minor downpour. The liquid tasted almost like pure water. The nymph appeared to be having the same effect on the tree as a medicinal leech has on human skin. As it impaled the bark with its proboscis, the 'anti-coagulant' that it injected ensured that there was a continuous flow of sap into the bubble-chamber in which it lay protected.

Writing the present book, I have frequently been reminded of the episode of the rain and the froghopper because of the loose parallel that can be drawn between the froghopper's ability to 'milk' the sap from the tree, and the milking of nectar from flowers by insects. The causes and the effects in the respective cases are, however, completely different. As far as I am aware, the froghopper does no good to the willow tree, although a willow tree probably never feels the sap rise through its channels more invigoratingly than when it is afflicted with these unwelcome guests. Nectar, on the other hand, is given away freely to the insects as a lure to the flower. The flower is the means to the plant's reproduction, so in this sense nectar is as much part of the life-blood of the plant as the sap itself is.

If you have ever tasted nectar in the wild, you will understand why it is so sought after by insects. Children often sample it by sucking at clover flowers. Adults tend to shy away from this practice, but if you really want to savour 'nectar-in-the-round' try crushing the spur of a garden columbine between your teeth. This dainty experience will also enable you to appreciate more fully the account in Chapter 1 which explains why certain robber bees are so eager to get at the pools of nectar contained in flower spurs that they dispense with normal formalities and simply bite straight into them.

This book has been written in the same spirit of 'test it and see'. I firmly believe that the natural world – and that can mean your own back garden or the local walk between villages – is a classroom. A lesson learned through your own eyes, no matter how simple, is always a revelation and is always accorded a place in memory. And, like a good investment, it awakens the imagination to the next opportunity. Some of my own glimpses of the workings of insects and flowers are contained in these pages, wrought out of journeys to Europe, the Mediterranean, North Africa and the Middle East, and no less from forays into the Cambridgeshire countryside. This is but a small section of the globe but big enough, I think, to give an idea of the range and complexity of the partnerships that exist between insects and flowers.

A honey-bee collects pollen from almond blossom.

7

THE FLOWER AS A FOOD SOURCE

More than a hundred million years ago the plant kingdom discovered that as far as its relationships with insects were concerned, fair exchange was no robbery. Insects could be duped into rendering a service, i.e. pollination, as long as they were offered a suitable reward. So began one of the longest partnerships in biological history and over the intervening millennia the beneficiaries have become ever more closely attached. In return for its services, the pollinating insect receives nectar and pollen, and plants have evolved every means to heighten the allurement that their flowers present. With the development of colour, plants even succeeded unwittingly in striking at the heart of man's perception of beauty. It was left to the German scientist Sprengel, nearly two centuries ago, to remind us that flowers were evolved for the benefit of their pollinators, not for human aesthetic appreciation.

ABOVE *Two solitary bees,* Eucera, *roosting on a geranium.*

OPPOSITE *Solitary bees curl up in a crown anemone.*

NECTAR AND NECTARIES

Flowers produce two of the most vital ingredients in an insect's diet: carbohydrate and protein. Obligatory flower feeders, such as bees and butterflies, receive virtually all their nourishment from this single source. Yet, unlike the milk provided for a new-born calf by its mother, the flower is not a perfect food. Neither nectar nor pollen are rich in fats, essential amino acids, vitamins, salts or the commonest of all dietary requirements, water. The great advantage of nectar, however, is that it is virtually pure sugar in the form of sucrose, fructose or glucose. This is the simplest of all energy sources, needs little digestion in the insect's gut and is available for use by the tissues almost as soon as it is eaten. Weight by weight, only fat contains more energy than sugar and since in any case, an insect can readily convert its dietary sugar into fat for long-term storage in the body, sugar is almost the ideal cover-all.

Nectar is only one of several natural sources of sugar. Ripe fruit is rich in fructose: the tart taste of unripe fruit is due to acids which form the precursors of the sugar. Fermentation of over-ripe fruit converts the sugars to alcohol. All three forms – fruit acids, fruit sugar and alcohol – are interconvertible and of equal nutritional value to insects that feed on fruit. Sugar is also locked up inside more complex molecules such as plant starches and cellulose. The sugar can be rendered from these in the insect's gut, provided it possesses the appropriate enzymes. It is even possible to extract sugar from a diet of pure wood and this happens in the larvae of wood-boring beetles and termites. In most cases the insect itself is deficient in the necessary enzymes, but help is at hand through the natural gut flora which digests the cellulose into simpler carbohydrates. Subsisting on a diet of wood can be a tedious process and wood-boring larvae often take several years to realize the fruits of their labour in the form of a perfect imago.

Nectar is the ideal food source for insects like bees, butterflies and moths that spend a lot of time on the wing and generate a high demand for energy. Or, to turn this statement on its head, nectar-feeding insects need to spend a lot of time flying between flowers simply to gain enough food to suffice for their needs. There is no knowing exactly where this chicken-and-egg race began and clearly, throughout evolution, a delicate balance must have been struck so that during the course of its travails the insect gains enough nectar from visiting a minimum number of flowers. This can only be achieved, on the part of the plant, by ensuring that its pollinator does not tarry too long over each individual flower. This represents the strategy; the tactic used is the obvious one: to limit the amount of nectar produced by each flower. Here we see a beautiful example of a game played between mutually benefiting partners by careful manipulation of their respective time-and-energy budgets. The plant needs to invest sufficient energy into each flower unit to make it worthwhile for the insect to spend time foraging on it, but not too much, otherwise it will have to limit the number of units that can be displayed since each plant can only work within its own energy budget. The insect, too, must deploy its time and energy wisely: having to make an excessive number of visits, over large distances, to over-thrifty species of plant would be counter-productive.

Another problem, from the plant's point of view, is that its products are too popular. The market for pollen and nectar includes many insects that are of no use whatever to the plant since they are ineffective pollinators. The plant needs to add a coda to its advertisement, equivalent to the 'no time-wasters, please' phrase used by second-hand car advertisers, but this particular innovation is evidently beyond normal evolutionary capabilities since there seems to be no known species of plant that selectively greets bees, butterflies and moths and does violence to unwelcome intruders. The very opposite may prove to be the case: arums physically entrap their pollinators, which are usually flies or beetles, and it is only later, as the flower wilts and after the pollinators have performed their service, that an escape route is made available to them.

If it is not practical to keep out the intruders the next best thing is to contrive mechanisms that conceal the nectaries so well that it is beyond the wit of all but a few insects to locate them. This is equivalent to putting your valuables in a safe-box. Many flowers owe their extraordinarily exotic appearance to this effort to discriminate between pollinators and non-pollinators. The 'keys' to the nectaries are the preserve of a few insects possessing the necessary attributes of size, strength and innate intelligence. This subject will be dealt with more fully in Chapter 3, but for the moment we can take as an example the toadflax shown in photograph **87** on page 84. In colour, shape and construction *Linaria triornithophora* bears the hallmarks of a bee-pollinated flower, down to the coloured boss in the flower throat to guide the bee's tongue, and the extended spur in which most of the nectar is stored. Toadflax, like the snapdragon to which it is closely related, has a mouth that is clamped under tension so only an insect strong enough to prise it open can possibly reach the nectar by legitimate means. Close examination of photograph **87** will show that crooks have also been at work on these particular flowers. Too impatient to get into the flower through the legitimate but laborious route, some bumble-bees have dishonoured the special relationship and bitten straight into the spur. This appears to be a case of gamekeeper turned poacher, and the flower is helpless to resist it. One such robber has been caught *in flagrante* in photograph **89**.

One way to ensure that nectar is preserved for the bona fide pollinators is to set up decoy nectaries on parts of the plant lying well away from the flowers. In the case of the common vetch *Vicia sativa*, tiny bowl-shaped extrafloral nectaries are located on the surface of the stipules subtending each flower stalk. These are assiduously tended by the black ant *Lasius niger* which patrols up and down the plant stem, taking a sip from each nectary in turn, and fending off vagrant flies and beetles that might be tempted to try their own luck on the plant. In this delicate security situation one presumes there must be some provision for the ant to switch off its aggression when the proper pollinator, i.e. a bee, approaches the flower. Alternatively, it is possible that the incoming bee simply regards the scuttling ant as a minor inconvenience.

THE ECONOMICS OF FORAGING

The impression that bees and nectar go together is well borne out by scientific measurements. It has been estimated that a honey-bee can only be sustained in flight if its muscles receive a supply of 10 mg of sugar every hour. But how much nectar can an average honey-bee collect during an hour's foraging? Clearly there has to be a surplus, otherwise the equation wouldn't make sense and the arrangement would soon be terminated. The exact amount varies between species of plants, depending on how much nectar is secreted by each flower and the speed at which the bee can harvest it. Most nectars have a concentration of between 20 and 60 per cent, equivalent to between 20 and 60 grams of sugar dissolved in 100 ml of water. Bees generally prefer nectars with concentrations in the range 30–50 per cent, but flowers with lower concentrations may be selected if the bee is in need of water.

Some idea of the 'micro-economics' of nectar foraging can be gained from a consideration of the relationship between bumble-bees and red clover *Trifolium pratense*. Apart from being one of the bee's favourite food sources, red clover is also widely grown as a fodder plant on account of its high protein and mineral content. Here, as in countless other cases, man and insect are harnessed together in a common cause. A single clover head contains numerous florets that can be foraged by a *Bombus* bee at rates of up to 40 per minute. However, each floret contains only a minute volume of nectar equivalent to about 0.02 mg of sugar. An industrious individual, working non-stop, should be capable of gathering nearly 50 mg of sugar per hour. This will allow for the 10 mg that might be needed for flight between flowers, and still leaves plenty to take home in the honey-stomach. These of course are extreme figures: no bee is ever *that* restless!

Nevertheless it is a truism that the life of the worker bee is amongst the most arduous of all animals and not surprisingly it is the younger, fitter individuals that must carry the burden of the task of foraging for nectar and pollen. Older workers spend more and more time at home in the hive or nest,

11

taking receipt of the goods from the incoming younger bees and delivering it to the inner chambers. Seated at our breakfast tables, it is interesting to reflect on the effort that has probably been expended in producing a standard 450 g pot of pure honey. This will have necessitated approximately 17,000 honey-bee foraging trips during which 10 million individual flowers will have been visited. An average foraging trip takes in about 500 flowers and lasts about 25 minutes. In all, therefore, 450 g of pure honey represents an investment of 7,000 bee-hours of labour. For the consumer, honey is indeed incredibly cheap at the price!

This statement would be endorsed by many animals in addition to ourselves, including bears and humming-birds, although nectar also has some unusual connoisseurs. Biting flies such as mosquitoes and the stable fly *Stomoxys*, as well as the common house-fly, can and are willing to survive on a pure sugar solution for days or even weeks. Stable flies, despite their reputation as blood suckers, have been known to spurn the offer of a naked human arm in favour of a clump of golden-rod *Solidago*. But there is a limit to their idiosyncrasies and they know when sugar is not enough. Both stable flies and mosquitoes need protein if they are to develop eggs successfully and in these circumstances they resort to their natural instinct for a blood meal.

Butterflies, moths and skippers fall next in line to bees as nectar feeders, at least in volume terms, although from the plant's viewpoint they are far less effective pollinators. Lepidoptera are said to prefer more dilute nectars to balance the high water loss incurred during foraging flight, but this seems to be a notion with no real foundation. There is no a priori reason why the water economy of flying moths and butterflies should differ from that of any other group of insects. Skippers are commonly seen on lucerne *Medicago sativa*, the vetch *Vicia cracca* and red clover *Trifolium pratense*, all of which have a nectar concentration of about 60 per cent. In this particular case, the preference for concentrated nectar may be related to the additional energy requirements of female skippers which, even after metamorphosis, contain only immature eggs that need a further period of time to grow and develop. Most butterflies can lay fertile eggs without feeding but an intake of nectar at least boosts fecundity, probably as a result of the traces of amino acids carried within it. However, this can provide at most a 'top-up'; the bulk of the protein for egg production has already been accumulated in the caterpillar stage. A few tropical butterflies, notably the heliconiids, imbibe pollen grains along with nectar and use the protein directly for egg production.

COSTS AND BENEFITS

The energy budget of a butterfly, like that of any other insect or indeed any other animal, is divided into three main components. The first is survival or maintenance energy, the amount of energy needed simply to allow the body to keep 'ticking over' when it is at rest. Next comes activity energy, used up during walking, flying and feeding. Finally there is a reproductive component which represents the insect's deliberate investment of hard-won energy reserves into the perpetuation of the species. Butterflies, at least those living in temperate countries, may also need to make provision for a fourth component: hibernation energy. Regular hibernators such as the red admiral *Vanessa atalanta*, small tortoiseshell *Aglais urticae* and peacock *Inachis io* may have to survive a winter lasting 150 days or more, at average temperatures hovering just above freezing. It takes at least seven days of regular feeding in the preceding autumn to lay down enough energy reserves to be able to survive this period. Most of this energy, derived from the nectar of late-flowering species such as ivy, is stored in the body cavity as fat and glycogen; only very limited amounts can be stored as free sugar in the blood.

It is worth spending a short time looking at the mechanics of nectar feeding as it might be experienced from the butterfly's point of view. First of all, imagine what it is like trying to suck up treacle through a drinking straw. It is not possible, you would collapse a lung before you succeed. Add 40 ml of water to 60 ml of treacle to make a 60 per cent solution of sugar. It is still impossible to drink it through the straw.

Where we fail, the butterfly triumphs because this is precisely what happens whenever a skipper sips concentrated nectar. The achievement is even more impressive considering that the food canal running along the skipper's proboscis is only 0.03 mm in diameter. In photograph **49** on page 57 the food canal of a diurnal moth can be clearly seen outlined against the semi-transparent wall of the proboscis. The movement of fluid along the canal is assisted by capillary action augmented by a pump located in the head at the base of the proboscis.

Curiously, butterflies and humming-birds have arrived at the same architectural solution to the problem of sipping sticky fluids. Down the length of a humming-bird's tongue runs a pair of grooves, of the same diameter as the butterfly's food canal, and capable of exerting the same capillary action. A bee's tongue is shorter but similar in design and its pump must develop a negative pressure of about one atmosphere to draw nectar along the canal. If nectar were only a little stickier than it actually is, it would probably be undrinkable.

Finally, what is the cost of nectar production to the plant itself? This cannot be measured purely in energy terms. In a sense, nectar is a 'loss-leader': it may attract the right exchange from a pollinator; on the other hand it may simply fall into the grasp of a non-pollinator. The penalty, in purely energy terms, is not unsupportable: flowering plants invest no more than 5 per cent of their total productivity in nectar, although the exact rate of secretion will vary with conditions. Day-flowering plants secrete less nectar at night; plants having nocturnal pollinators secrete less during the day. In hot weather, the rate of evaporation may exceed the rate of secretion so that sugar crystallizes out on the surface of the nectaries. If the flower is not visited by insects, nectar may accumulate in globules as shown in photographs **10** and **11** on pages 20 and 21. All in all, it is probable that nectar secretion affects the water economy of the plant at least as much as its energy economy and in circumstances of severe water deficit this may become critical. This subject is taken up again in greater detail in the final chapter.

1. The honey-bee *Apis mellifera* is the most familiar species in the partnership between insects and flowers. Here seen swarming in springtime, it is the complex social life and provident behaviour of the honey-bee that has enabled it to achieve dominance in the flower market.

2, 3. Flowers have their enemies as well as friends among insects. These include destructive caterpillars, such as those of hawk-moths here seen feeding on flowers of the shrub tobacco *Nicotiana glauca* (**2**) and spurge *Euphorbia dendroides* (**3**). Both these plants contain poisonous chemicals which are extracted by the caterpillars to render themselves distasteful to predators. When attacked, the caterpillars protect themselves by releasing a noxious, bile-like fluid from the mouth which quickly spreads over the head and upper parts of the body

4–6. The chafer beetle *Oxythyrea* destroys a wide range of flowers, even when they are in bud. Here (4) it has just landed on an alkanet *Anchusa strigosa* and, undeterred by the tuft of hairs surrounding the mouth of the flower, has begun to chew its way inside. Soon (5) it will have burrowed its way down to the most nutritious part of the flower, the ovaries. Similarly (6), the beetle has ignored the less tasty petals of a sage flower *Salvia argentea* and is busy chewing the ovaries.

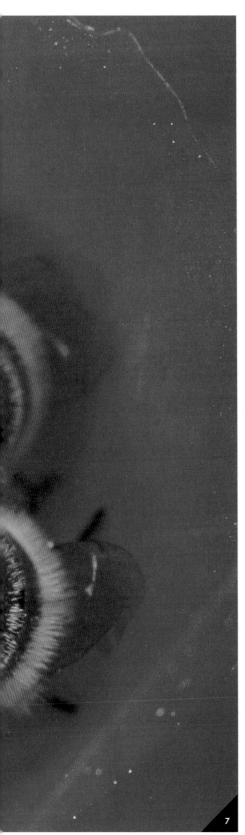

7, 8. In addition to food, some flowers also offer overnight accommodation to their pollinators. A party of the solitary bee *Eucera* has found the cup-like flower of crown anemone *Anemone coronaria* particularly amenable (**7**) and a pair of *Eucera* are here seen curled up for the night inside a *Geranium* flower (**8**). The main pollinator of *A. coronaria* is the beetle *Amphicoma* illustrated in **108** on page 110 but there seems little doubt that *Eucera* contributes. In return, not only does the bee receive shelter and protection overnight, but also it benefits from the increased humidity created as the flower closes up for the night.

9. Nectar is almost pure sugar, and is therefore a concentrated source of energy for insects. Once eaten it is rapidly absorbed and digested and ready for use by the tissues. The flight muscle of a hovering bee, such as the one shown here, burns up sugar at a faster rate, weight for weight, than any other known tissue.

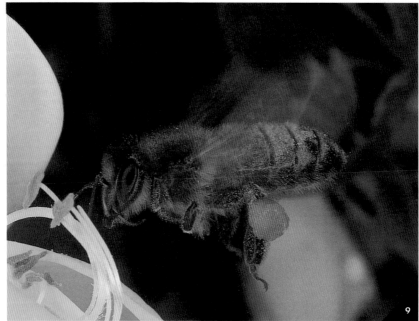

10, 11. The nectaries of flowers are usually located at the base of the petals. Photograph **10** shows the view that a bee would have after crawling inside the inverted mouth of an imperial lily flower *Fritillaria imperialis*. Its attention would be drawn to the white nectaries, highlighted against the dark interior, each bearing a teardrop shaped blob of nectar. *Fritillaria* nectar is quite watery and probably has a concentration of about 30 per cent sugar. The globules of nectar on the surface of these *Fatshedera* flowers (**11**) is much stickier, and its sugar is more concentrated. The strongest nectars have sugar concentrations of up to 60 per cent. Anything more concentrated and therefore even 'stickier', would be almost undrinkable through the narrow proboscis of specialist nectar-feeders like bees and butterflies.

11

12–15. Common ivy *Hedera ilex* is closely related to the *Fatshedera* of **11** on page 21 and has an almost identical flower, which is simple in construction and requires no special adaptations on the part of its pollinators to obtain its readily available nectar. *Hedera* is a late flowerer and is on this account especially beneficial to insects living in temperate countries and preparing for winter hibernation. Amongst its regular clientele are long-tongued insects such as honey-bee (**12**) and red admiral butterfly *Vanessa atalanta* (**13**) and less extremely adapted species like the social wasp *Vespula* (**14**) and the blowfly *Calliphora* (**15**). Other insects commonly seen feeding on *Hedera* in autumn are the dronefly *Eristalis*, the comma butterfly *Polygonia c-album* and greenbottle flies *Lucilia*.

16, 17. This section of a *Fuchsia* flower shows the green conical nectaries at the bottom of the long corolla tube (**16**). This is a young flower and the petals are not yet fully opened, nor have the anthers ripened. *Fuchsia* is normally pollinated by birds in its native environment but the cultivated variety shown in **17** is being visited by a bee. To judge by the pollen on its body the bee has already made an unsuccessful attempt to reach the nectaries via the mouth of the corolla but it has now changed tack and is inserting its tongue between the petal bases.

18, 19. The nectaries of *Geranium* (18) lie between the bases of the petals: the access holes to two of the five nectaries can be identified in the photograph. The *Geranium* flower is orientated almost vertically but visiting bees gain a purchase by clinging to the staminal column. In **19** a solitary bee *Osmia*, identifiable by the fox-coloured pollen basket on the underside of the abdomen, is using this purchase to direct its tongue into one of the access holes. Although there are five nectaries, most bees seem to probe only the upper three, neglecting the lower two since it is evidently too time consuming (or the bees are simply in too much of a hurry!) to position their bodies upside down on the staminal column.

2

THE MARKET
FOR
POLLEN

P OLLEN IS THE male gamete of the flower, although it would be more accurate to describe it as the vehicle in which the gamete is conveyed since the bulk of a pollen grain consists of the tough, sculpted case enclosing the gamete which is itself physically minuscule. Almost universally among plants and animals it is the sperm that is motile and must propel itself, or be carried, to the immobile female gamete or egg. Because they have to do the searching, as it were, male gametes are produced in prodigious numbers in order to increase the likelihood of encounter with the opposite gamete. They normally reach the egg by propelling themselves through a watery medium, which may be seminal fluid in the case of animals that copulate; the thin film of moisture covering the leaves of mosses or liverworts; or a large body of water, as in the external fertilization of the eggs of fish and sea-urchins.

ABOVE *Bees visit poppies for pollen, not nectar.*

OPPOSITE *A long-horn beetle browsing on thistle pollen.*

Pollen dispersal

Flowering plants are somewhat exceptional in their choice of a dry sperm which lacks self-propulsion. As a result, pollen must be conveyed passively to the receptive stigma of the flower, either by the wind, which constitutes an entirely random delivery service, or through the more directed agency of pollinators, usually insects or birds. A few water plants such as hornwort *Ceratophyllum* and the pondweeds *Elodea* and *Potamogeton* retain the services of water but even then the transfer takes place within the surface film.

In biological terms, eggs are infinitely more valuable than sperms. Over 99.99 per cent of sperms have rendered themselves redundant through profligacy even though they carry the same genetic potential as the egg. Every cell of an adult organism belongs to a lineage derived from the union of two gametes, each of which provides exactly half the complement required for a full set of chromosomes. Only the egg, however, contains enough energy reserves from which a whole miniature organism can be produced: a plant seed is a fertilized egg that has been bolstered with food reserves to provide the makings of a perfect seedling. It needs only a little water from the outside world to enable it to get on with the job. A fertilized bird's egg is even more self-sufficient; it contains everything that is required to make a chick, bar some warmth provided by the mother.

Every egg produced by a female plant or animal represents a considerable investment of materials and energy and that is why, compared to sperms, so few of them are manufactured. A tree for example may liberate a million or more pollen grains for every ovule produced in a flower. This figure itself testifies to the colossal wastage inherent in a wind-pollination system since, of course, only one pollen grain is needed to fertilize an ovule. But the flower has to work hard even to achieve a million to one success rate: wind-pollinated flowers usually need elongated, feathery stigmas to increase the efficiency with which they trawl the air for pollen grains. The principle is the same as that seen in the huge, plumose antennae of some moths which filter the air, not for pollen, but for highly diluted aphrodisiac chemicals which may have been liberated from a member of the opposite sex located several kilometres away.

Using insects to carry pollen directly from one flower to another is obviously more economical than relying on the vicissitudes of the wind, but insect pollinators vary greatly in their efficiency. Every insect that visits flowers is a potential pollinator, but very few realize this role effectively either because their visits are not frequent enough, or because their bodies are poorly designed for picking up pollen for transfer to the next flower. Infrequent or irregular visitors are not reliable, even if they carry pollen. The mosquitoes and stable flies referred to in the previous chapter, for example, only take an active interest in flowers as a diversion from their normal diet. The rust-coloured dung-fly *Scopeuma* may be seen in early springtime breaking its winter fast on lesser celandine *Ranunculus ficaria*, and may contribute to the pollination of this species, but as the season progresses it reverts to its more natural carnivorous habit, and then simply uses flower heads as convenient places for ambushing other insects. A similar example can be seen in photograph **34** showing the empid fly *Empis tessellata* drinking nectar from a *Convolvulus* flower. If you look carefully, you will see pollen grains stuck to the hairs of its thorax. But *Empis*, like *Scopeuma*, is primarily a carnivorous fly and its attachment to flowers is not constant enough to make it a dependable pollinator.

It is a fact that plants require loyalty from their insect partners; although few insect pollinators, even bees, confine their attentions to a single or even a small number of species of plant, they do at least make frequent visits between members of the same species. Having discovered during the course of its foraging a highly rewarding species of flower, it is therefore in the bee's interests to continue to exploit this rich vein for as long as it lasts. Flower constancy of this level requires the kind of associative learning capability that the majority of flies, beetles and bugs do not possess.

THE QUALITIES OF A GOOD POLLINATOR

A good pollinator must display the right behavioural patterns but it must also come equipped with the right degree of hairiness. These two qualities do not always coincide in the same insect. Parasitic wasps such as pompilids, sphecids and ichneumonids are amongst the most assiduous nectar seekers and in terms of frequency of flower visits, they could probably rank with many bees. But they are almost completely naked and consequently they make poor pollen couriers. The social vespids have rather hairier thoraxes and are known to pollinate figworts of the genus *Scrophularia*. In addition, I have seen *Vespula vulgaris* crawling from one flower to another of policeman's helmet *Impatiens glandulifera*, its body covered in white sticky pollen. Butterflies and moths, on the face of it, seem to have all the right qualities: total dependence on flowers as a food source, a hirsute body, and an elongated tongue giving them access, at least in theory, to specialized flowers. In practice, however, butterflies usually prefer quite shallow flowers, such as the florets on the discs of composites. A typical butterfly flower such as *Buddleia* has a narrow tubular shape and a flat rim on which the butterfly can alight. Butterflies are rarely called upon to probe so deeply into a flower that their heads come into physical contact with the floral parts, particularly the stamens, and this explains, at least in part, their relative ineffectiveness as pollinators. Moths, in contrast, probe deep-throated flowers such as campions, catchflies and honeysuckle and are the normal pollinators of these flowers.

Some species of fly are almost as specialized for flower-feeding as bees, both in their habits and the manner in which their bodies are constructed. The Nemestrinidae and Bombyliidae, examples of which are shown in photographs **67** and **68** on pages 68 and 69 have extremely hairy bodies and long, rapier-like proboscis. Unlike the equally exaggerated instrument of the empid flies, the bombyliid proboscis is used exclusively for nectar feeding. The proboscis of the common hoverfly *Rhingia*, the subject of photographs **65** and **66** on pages 66 and 67, is almost a centimetre long and it can reach into quite deep, tubular flowers that are normally pollinated by bees. This particular fly is known to manipulate the balance of nectar and pollen in its diet, according to its reproductive condition. During the breeding season, when the demand for protein is high, the pollen intake is augmented. When the breeding season is over, and eggs are no longer being produced, the fly reverts to its normal diet of pure nectar.

More than any other group of insects, bees are both the driving force and the product of the floral market. In evolutionary time, flowering plants or angiosperms pre-date bees although exactly when and how this most specialized group of plants arose is still, in the words of Charles Darwin, 'an abominable mystery'. A kind of co-operative arrangement between insects and plants probably existed long before the advent of angiosperms: botanists cite the pollination of modern cycads by insects in support of this theory. Somewhat unexpectedly, the most primitive group to have lent their sevices in this manner is thought to be beetles and the fossil record confirms that beetles were abundant during the Jurassic and Cretaceous periods when angiosperms began to radiate. Apparently the beetles' methods were not very sophisticated since they simply scrabbled about amongst the stamens, drenching their bodies in pollen, which they then chewed with their strong mandibles.

The lapse of a hundred million years has done little to improve the beetles' habits: 'mess and soil' pollination, as it is referred to, remains the basic floral technique of modern beetles. But it is nevertheless effective since beetles are the main pollinators of many simple types of flower, as will be seen in the next chapter. Amongst the beetles seen most commonly on flowers are the soldier beetles (Cantharidae) one of which is shown in photograph **52** on page 59 and longhorn beetles (Cerambycidae). These are mainly interested in the nectar that the flower has to offer, but the flower beetle *Oedemera* featured in photographs **28** (page 38) and **32** (page 40) and the tiny, black *Meligethes* that clusters around stamens of daffodils, dandelions and dog's mercury *Mercurialis perennis* in springtime, are both specialized pollen chewers.

Unfortunately there are also beetles that destroy flowers; one of the worst offenders is *Oxythyrea* shown in photographs **4–6** (pages 16–17). At first sight it could easily be mistaken for a harmless pollinator, poking its head inquisitively into the mouth of the flower. Before long, however, it will be seen to have nibbled its way down to the bottom of the corolla tube where it is now ready to dispatch the ovaries, the richest energy source in the whole flower head.

One of the things that distinguishes bees from most other insects is their providence. They collect food, not to consume it on the spot, but to take it away and store it for future use. This is in the flower's interests as well, because it guarantees that once a bee is embarked on a foraging run, it will not cease until it has visited a large number of flowers. If it likes the location, it will come back, time and time again. Even though the collector will endeavour to stow away as much of its hard-won gains as possible in its pollen baskets, there will always be some pollen that eludes its grooming movements and remains available for transfer to the stigmas of consecutive flowers. In any case, pollen rapidly loses viability once it has been compacted into the pollen baskets, so the plant accepts these relatively huge losses in the interests of securing success for the few.

Pollen baskets come in several varieties, in addition to the familiar structures seen on the hind legs of honey-bees and bumble-bees. The hairs on the underside of the abdomen of the solitary bees *Osmia*, *Megachile* and *Anthidium* form a dense pile which acts as a pollen brush (see photographs **19** on page 25, and **40** and **41** on pages 45 and 46). *Andrena* carries pollen home on the main joint of the hind leg but both the abdomen and thorax are densely furry as can be seen in photograph **55** (page 61). A regular grooming routine ensures the maximum retrieval of pollen grains. The front legs are deployed in combing through the head and the front of the thorax, making use of a notched antennal cleaner, which can be seen in photograph **48** on page 57. These same legs are surprisingly dextrous: photograph **26** (page 36) shows a honey-bee holding individual anthers like ice-cream cornets whilst using its mandibles to nibble at the pollen. The middle legs rake through the pelage on the sides and top of the abdomen. Especially sticky pollen could clog up the wing bases, interfering with flight, and photograph **38** on page 44 shows a honey-bee assiduously grooming its wings after emerging from the funnel of a *Convolvulus* flower covered in granules. Finally, with a series of scraping movements between the legs, pollen is transferred either to the abdominal pollen brush, or, in the case of the honey-bee, first on to the inner side of one hind leg then the outer side of the opposite leg. The pollen mass is then held in place on the leg by a single bristle spearing it through its centre (see photographs **25** and **26** on pages 35 and 36).

Two parts of a bee's body are particularly difficult to groom: the areas beneath the head and directly in the middle of the back. Some flowers exploit this chink in the insect's armour by targeting, with their stamens, precisely these two areas. A good example is seen in certain Leguminosae, such as brooms, which are pollinated by an explosive mechanism described in the next chapter. The result is that as the insect alights on the flower the stamens whip themselves over its head on to its back bringing the anthers into contact with the 'safe site'.

20. Flowers are not all designed to attract insects. Grasses, sedges, rushes and stinging nettles rely on wind pollination as do many species of tree with inconspicuous flowers. The cotton-sedge *Eriophorum angustifolium* shown here looks conspicuous enough on account of its tufts of silky hairs but these are all that is left of the petals and sepals.

20

21. Trees of temperate climates such as the catkin-bearing birch, alder, and beech flower very early in the year, in some cases long before insects have emerged from winter hibernation. Ash, sycamore, maple and the elm shown here have more conventional flowers but they, too, are largely wind-pollinated. Elm is already in blossom by early February, even when snow is lying on the ground. Notice the numerous stamens of these elm flowers liberating clouds of pollen grains each measuring only 0.03 mm or so in diameter. In still air these will sink steadily at a rate of a few centimetres per second and theoretically it could take up to a quarter of an hour for grains from the highest flowers on an elm tree to reach the ground. This would be wasted pollen but the flower is so constructed that even though the stamens may have cast their pollen, it collects within the flower until jostled by a suitably powerful breeze.

22, 23. The corn poppy *Papaver rhoeas* (**22**) is a typical pollen flower: it is showy, in order to attract a wide variety of insects, it is easy to get into, and it contains a forest of stamens in which insects can scrabble about, covering themselves in pollen. Common visitors to poppies which indulge in 'mess and soil' pollination of this kind include small solitary bees, honey-bees and beetles. In this case (**23**) a speckled bush-cricket *Leptophyes punctatissima* is nibbling protein grains that have become stuck to its feet.

24, 25. Bramble *Rubus fruticosus* flowers soon after hawthorn and at about the same time as elder and dogrose. Like poppy it is mainly a pollen flower and it is popular with honeybees, bumble-bees, drone flies and longhorn beetles. Honey-bees (**24**) probe the central tuft of stigmas for nectar while at the same time vigorously combing the stamens with their legs. Pollen and nectar are then kneaded into a dough which is transferred to the pollen baskets on the outer side of the tibia of the hind legs. The transfer route is quite complicated and involves first of all the pollen being collected on to a brush on the inner side of the hind tarsus. The left tarsal brush can be seen in **25** showing a honey-bee on another pollen flower, a rock-rose *Helianthemum*. From the tarsal brush the pollen is then squeezed through the tibio-tarsal joint on to the outside of the tibia, where it can now be seen, for instance, in **26** on page 36.

24

26, 27. A honey-bee balances
delicately on the pin-cushion of
stamens of an almond flower *Prunus
dulcis* (**26**). It is using its front feet to
manipulate the stamens towards its
mouth where it nibbles them with its
mandibles. With similar dexterity a
hoverfly *Helophilus* grasps a stamen of
an ivy flower between its front feet
and dabs it with its proboscis (**27**).

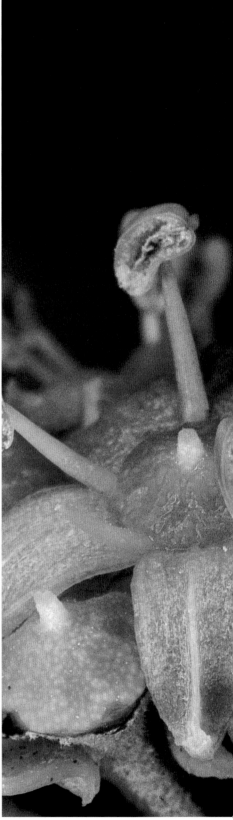

28, 29. A flower beetle *Oedemera nobilis* clings to the staminal column of a musk mallow flower *Malva moschata* and browses on the pollen from the freshly opened anthers (**28**). *Malva* pollen is very sticky and each 'grain' adhering to the beetle's body is an accretion of many individual pollen grains. Musk mallow (**29**) is one of the numerous species of wild flowers found growing in sub-alpine meadows of the Picos Europa mountains in northern Spain. The abundance of wild flowers, and their dependent insects, in this region is a direct result of the use of traditional mowing methods by the local farming community. The long staminal column of musk mallow serves a dual purpose, first by providing a pole to which visiting bees can cling while sipping nectar from the base of the petals, and second by placing the stamens beyond the reach of non-pollinators. The specialist pollen feeder *Oedemera* is deft enough to overcome this obstacle (**28**).

30. The elongated head, jaws and prothorax of longhorn beetles such as the one shown here feeding on a thistle head are adaptations for nudging between closely arranged florets in search of pollen. Beetles are considered to be the most primitive flower pollinators and during the Jurassic and Cretaceous periods of geological history when flowering plants first evolved they even dominated in this role over bees. Click beetles, rove beetles, chrysomelid and longhorn beetles are the ones most commonly seen feeding on flowers. Some chafers are pollen feeders (see **108** on page 110) whilst others destroy the whole flower (see **4–6** on pages 16–17).

29

30

31–34. The bindweed *Convolvulus althaeoides* (31) advertises itself to a wide range of prospective pollinators although only long-tongued insects can reach its nectaries. Here (32) the flower beetle *Oedemera nobilis* feeds on pollen from the anthers of *Convolvulus arvensis* and in the related bell-bine *Calystegia sepium* (33) a hoverfly is using the stigmas as a foothold for mopping up stray pollen grains. *Convolvulus* and *Calystegia* are examples of 'revolver-flowers' because the nectaries are hidden at the ends of a series of five narrow passages positioned like the chambers of a revolver at the base of the stamens. Only a long-tongued insect, such as this *Empis tessellata* fly (34) can reach the nectar, and it must move round from one chamber to the next in sequence, brushing against the stamens as it goes. This is the flower's way of detaining the insect, and in so doing maximizing its chances of pollination.

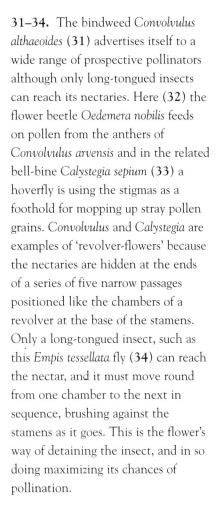

35–37. Most plants avoid self-pollination by altering the timing between the maturation of the anthers and the stigmas. In Campanulaceae, Labiatae, Caryophyllaceae, Compositae and Papilionaceae the anthers usually mature first, a condition known as protandry (literally, anthers first). These three photographs show successive stages in the development of the male and female parts in Malvaceae as exemplified by *Lavatera* (**35, 36**) and *Malva sylvestris* (**37**). In **35** the flower is in the male phase and is being pollinated by a brown bee *Bombus pascuorum*. This is still a young flower and only the topmost stamens have released their pollen grains from the thin membranes enveloping them. The bee, which is aiming its proboscis between the petal bases to the nectaries on the obverse side, is carrying a few sticky pollen balls from a previous visit to another, older *Lavatera* flower. In **36** the anthers have now all dehisced and some of their pollen has become attached to the top of the head and thorax of the visiting honey-bee. As yet, there is still no sign of the stigmas. In **37** an empid fly is probing amongst the bases of the stamens which are now old and reflexed, while the slightly feathery stigmas are now protruding from the end of the staminal column like a bunch of fibre optic cables, and are making contact with the underside of the fly. A few pollen grains are attached to the stigmas but they do not appear to have come from the fly, whose body is not hairy enough to make it a very effective pollinator.

43

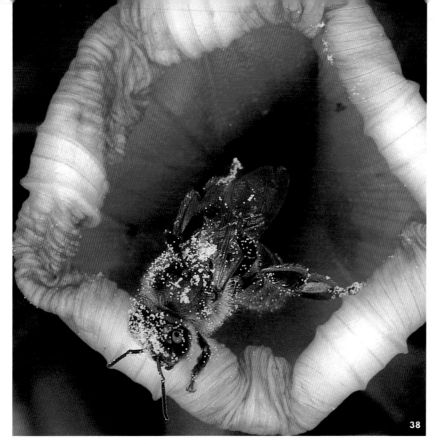

38. Pollen clings to the hairs of insects, particularly bees, by a combination of physical stickiness, and electrostatic charge. This honey-bee emerging from the rolled-up corolla of a *Convolvulus* flower is covered in sticky pollen. Note how it is using both of its middle legs to groom the pollen away from its back and from the bases of the wings. The most difficult parts to groom are the centre of the back and the area just beneath the proboscis. These 'safe sites' are particularly important from the plant's point of view since it is only ungroomed pollen that has a chance of being transferred to the stigma of the next flower.

39–41. Solitary bees often collect pollen on specialized areas of pubescence called pollen brushes. In *Andrena* (**39**) the densely hairy thorax is particularly important whilst *Megachile* (**40**) and *Anthidium* (**41** on page 46) have pollen brushes on the underside of the abdomen. The underside of the *Megachile* bee, here seen in the typical up-ended posture that it adopts when visiting thistle flowers, is luminous with yellow pollen grains. Note, incidentally, that thistle pollen is white not yellow, so the grains carried by the bee must have another provenance. The stiff hairs making up the abdominal pollen basket of *Anthidium* are clearly seen in this flying specimen (**41**), but note that its upper surface is entirely naked. This is probably to ensure that the prominent black-and-yellow warning coloration of the bee's body is not obscured by hairs.

40

41

42. The hoverfly *Episyrphus balteatus* shown here is dabbing with its proboscis the brown powdery material that has collected on the petals of a white campion *Silene alba*. The powder has come from the stamens that can be seen poking through the mouth of the corolla, but it is not pollen. The flower is in fact a female which has become masculinized as a result of infection by a fungal smut which is here seen attacking the anthers. The powder consists of millions of fungal spores which are evidently as nutritious as the pollen that they have replaced. Unwittingly, the hoverfly serves as a vector for the smut by carrying spores to the next flower that it visits.

42

47

FLOWER TYPES
AND THEIR
POLLINATION

ONE OF THE great pleasures of travelling through Mediterranean countries in springtime is the sight of meadows, hillsides, roadside verges and the edges of cornfields covered in flowers. One is struck not only by their colour but also their variety; if for a moment one were a bee, the choice could seem overwhelming. Yet a little time spent in observation might suggest otherwise. Whereas some types of flower appear to be popular with all the insects, others have a more specialized clientele. To illustrate the point, we can take two flowers from opposite ends of the spectrum. Photograph **102** on page 106 shows a typical sight that could be seen almost anywhere around the Mediterranean from late March, through April until May; a sward of luminous crown daisies *Chrysanthemum coronarium*. This beautiful plant enlivens any location in which it grows and turns waste land in towns into pools of golden sunshine. Another of its favourite habitats, where it is almost guaranteed to be free of disturbance is within the perimeter of designated archaeological sites. To a naturalist these protected sites are, by default, perfect nature reserves.

ABOVE *A bee deftly targets the narrow throat of* Pentaglottis.
OPPOSITE *A peacock butterfly probing a dandelion for nectar.*

A short time spent strolling amongst the crown daisies will confirm that they are attended by a wide variety of insects; in fact they provide a convenient cross-section of the local insect community. Caterpillars, flies, capsid bugs, beetles, green lacewings, bush-crickets and small solitary bees are among the many types that frequent it. If you now abandon the open sunshine and take to the shadier hillsides, especially into pine woods, you may find yourself having to step carefully over orchids. The air, faintly scented with the smell of turpentine, is still with barely an insect to be heard or seen. Certainly not on the orchids. Orchids provide the supreme example of a flower specialized for pollination by a narrow portfolio of insects, usually bees. In structure, colour and habitat orchids could hardly be more distinct from the crown daisy. In 'sales technique', too, these two belong to different schools. The crown daisy aims for mass appeal, organizing itself into broad stands that radiate bright colour towards not only insects on the ground but, more importantly, insects in the air. Like migrating geese that dive towards light reflected from bodies of water below, passing beetles, bugs and flies are drawn downwards to the lakes of colour. In both cases, light sends a promise of food. Once the beetle has drawn close it finds a mass of tightly spaced flower heads that are large, round, flat-topped and amenable for landing. Its crude, relatively uncoordinated manner of flight is no disadvantage in these circumstances: in truth it can hardly fail to hit a target. It alights to the sight of scores of small, shallow florets, uniformly arranged over the surface of the disc. Little effort or imagination is needed to extract the nectar: the only problem is the likelihood of another insect arriving to claim its share of the provender.

Imagine an insect of such narrow capabilities confronted with a much more complicated flower such as the broom *Cytisus* shown in photograph **76** on page 75. Two of the main markers to a reward of nectar or pollen are obviously present: colour and a sweet scent of vanilla. But only an insect with the appropriate tactile and visual understanding of the flower can find its way to the nectaries. By acquiring a design that discriminates against all but a special group of insects, which are highly efficient in their pollination, the broom plant can be more economical with its resources. In the extreme case of explosive pollination, described in detail later on, a single visit by a bee is sufficient to fulfil the role of the flower. This high-risk strategy only works because of the unique matching that exists between the partners. The partnership is based on the appropriate balancing of investment and reward. The reward is in the form of nectar and pollen which, barring a few cases of thievery, are the preserve of the specialized pollinator. The investment is the pollinator's time and energy. The gamble that the flower takes is that the bee has spent so much time and energy learning to crack the code, enabling it to bypass the flower's defences and gain access to the nectaries, that it needs to make its investment pay dividends.

During its relatively brief lifespan, a worker bee will have made thousands of decisions, on a trial-and-error basis, culminating in an understanding of many kinds of floral devices. In addition, she will have learnt to memorize trails between the hive and different flower sites, and how to communicate this information via body language to other workers in the hive. Yet paradoxically neither this capacity for learning and storing information, nor the possession of a sting, prepares her for the assaults of spiders that specialize in preying on insects foraging on flowers. Spiders like *Misumena* and *Thomisus* are frequently only a fraction of the size of their prey, but they overpower them with ease. A bee's intelligence does not necessarily extend to effective self-preservation.

In Chapter 1 a case was mentioned of certain types of bee that use their native intelligence to create short cuts to the nectaries of complex flowers. When interference of this kind leads to physical damage to the flower it is referred to as robbery, as opposed to thievery which only involves purloining the nectar by devious means. A robber is shown at work on a *Daboecia* flower in photograph **83** (page 80), whilst the bona fide pollinator of this plant, the bumble-bee *Bombus pascuorum* is portrayed legitimately going about its task in photograph **82** (page 80).

It is difficult for a flower to keep out all potential thieves since even specialist pollinators need some visual cues to lead them to the mouth of the corolla, and this may give the game away to others. Nectar guides provide a convenient way of directing the pollinator's tongue towards the mouth: examples are the radiating lines on the petals of the *Geranium* in photograph **18** on page 24, the bright orange throat boss of the toadflax *Linaria triornithophora* (photograph **87** on page 84) and the blotched lip of the foxglove *Digitalis purpurea* (photograph **85** on page 82). Nectar guides may also reflect strongly in the UV part of the light spectrum, to which bees and other insects are very sensitive. Even if unwanted visitors manage to read the nectar guides they still have to get past the flower's defences. In the case of the *Geranium* the central column of stamens, which bees negotiate with adroitness (see photograph **19** on page 25), presents a hinderance to less agile insects. The lip of the foxglove bears, in addition to the give-away spots, a series of backwardly directed hairs that give purchase to the bee but deter smaller insects.

FLOWER DESIGNS AND FORAGING STRATEGIES

A flower that is too elaborate in design becomes self-defeating since it may deter the pollinator as well as the thieves. We saw in Chapter 1 that a bee runs its life on a time-and-energy budget. If an inordinate amount of time is needed to gain access to the flower, in comparison with the reward that is offered, the pollinator will soon withdraw its services. On top of the access time, the insect also needs time to mop up the nectar. Generally, long-tongued bees are both nimbler about the flower and speedier in their drinking than short-tongued varieties. Hence they are more economical and can work a greater number of flowers in a given time.

Typically a *Geranium* flower will be processed by a bumble-bee such as *Bombus pratorum* in less than three seconds; this may involve inserting the tongue into five different nectaries, each located between the bases of neighbouring petals. Often the bee is working 'blind' in the sense that it must snake its tongue into nectaries that it cannot see. It can only

succeed in this task in the light of knowledge gained from experience, coupled with a natural tactile sense. The tiny florets of a clover head can each be worked in less than a second by a well-practised bee but kidney vetch, *Anthyllis vulneraria*, presents more of a barrier because the contracted throat of each floret demands a more precise placing of the bee's tongue. The overall time and energy balance sheet must also take into account time spent on the wing. For example, if a bee has to fly one kilometre to reach a patch of red clover the time required is approximately seven minutes, equivalent to time that could have been spent working approximately 270 clover florets.

It should be pointed out that successful foraging is more than a series of happy accidents. A bee learns to associate rewarding flower sites with particular landmarks. Acquiring such a mental map means that the bee must have at least short-term memory. It would be wrong to overstate the complexity of these 'route maps': they usually follow a well-defined and easily recognizable line of plants, a technique known as 'trap-lining'. Once a rewarding stand of flowers has been located, and always bearing in mind the time-and-energy saving imperative, it makes sense to forage in such a way that each flower is not visited more than once. In practice it is unlikely that the bee will be the only individual to have discovered this productive site and whatever strategy it sets for itself it is likely to keep bumping into others. Overall, however, the simplest plan seems to be to keep turning right or left after each flower, with roughly equal likelihood, so that a zigzag line is followed. Every now and then the bee can completely switch tack and start working the patch along a different line.

When foraging compound blossoms such as foxgloves, *Aconitum* or willow-herb *Epilobium* which consist of numerous flowers arranged in a vertical succession, a strategy is also needed but it must be simple. In such spiked inflorescences the arrangement is usually chronological with the youngest florets being those nearest the top. The bee's method is to work from the bottom upwards and one of the simplest ways to see this in action is to spend time watching bees working the common garden 'red hot poker' *Kniphofia*. As the bee steadily ascends the spike, from one floret to another, it finds that the quantity of nectar gradually diminishes but its concentration increases. So the reward remains the same and the bee does not feel cheated visiting the younger florets. This is just as well because the plant is relying on the bee proceeding from the youngest flowers of one individual to the oldest flowers of the next. In temperate plant families it is common for the male reproductive parts, the stamens, to mature before the ovaries in order to prevent self-pollination. Therefore in the case of spiked flowers the bee will transfer pollen from an upper, younger flower of one plant to the stigmas of a lower, older flower of the next plant.

43

43, 44. Umbelliferous plants like wild angelica *Angelica sylvestris* (**43**) and giant fennel *Ferula communis* (**44**) produce multitudes of tiny flowers arranged into larger, compound units or umbels. This is a subtle device enabling the plant to retain very large numbers of small flowers, each capable of setting seed, whilst at the same time remaining showy and attractive to insects. Umbellifers are usually coloured white, pink or yellow. Their combination of colour, simple structure and availability makes them attractive to a wide range of generalized pollinators such as beetles, flies, short-tongued bees, ichneumon flies, social wasps belonging to the genera *Vespa* and *Vespula* and parasitic wasps such as the spider-hunting *Sceliphron*.

45. Sunflower *Helianthus annuus* achieves the same ends as the umbellifers but uses a different means. In this case, as in most composites, the numerous individual florets are arranged on a disc ('disc florets') and are surrounded by 'ray florets' with much larger, showier petals. It is the latter that do most of the 'advertising' through their colour and size. Although it is a native plant of South Africa, sunflower has been cultivated in the warmer parts of Europe since the 16th century. Even so, it remains an alien as far as many European insects are concerned and its principal customer during the daytime is the highly adaptable honey-bee. At night, however, it also receives visits from lacewings, earwigs and grasshoppers. Notice how in this photograph the bees are concentrating their search towards the peripheral florets of the disc. This is because the disc florets open in a centripetal manner, the older flowers being on the outside. The inner flowers have not yet opened.

46. *Rudbeckia laciniata* is another large composite, with a conical rather than a flat receptacle, which is surrounded by a single row of large, sterile ray florets. In this specimen, only the outermost circles of disc florets have opened and this is where the bee is locating the nectar. The transition from the more central flower buds, each showing a five-pointed rudiment of the petals, to the fully opened florets on the outside, can easily be recognized.

45

47

47–49. A small skipper *Thymelicus flavus* (**47**), a bumble-bee *Bombus terrestris* (**48**) and a noctuid moth (**49**) sip nectar from a ragwort flower *Senecio jacobaea*. All three are very long-tongued insects and as a consequence of this their eyes do not come very close to the florets on which they are feeding. Insects are generally so myopic that it is likely that these three insects can only just distinguish one floret from its neighbours visually. The ragwort being visited by the skipper is a young flower with only the outer disc florets open. That visited by the bee is older and all the disc florets are open and displaying two-pronged stigmas. Careful examination of **47** shows that in this particular flower the youngest, central florets are in the male phase because they bear stamens but, as yet, no stigmas. This is the normal condition in composites; the stamens mature in advance of the ovaries (protandry). The food canal of the noctuid moth is seen running down the centre of the proboscis. Each half of the proboscis consists of an elongated maxilla with a gutter running along its inner edge. When the two halves are fused together they form the food canal. At 0.03mm in diameter, it is just wide enough to transport pollen grains in suspension.

48

49

57

50

50, 51. In another composite, the dandelion *Taraxacum officinale* shown in **50** being visited by a peacock butterfly *Inachis io*, the flower head consists entirely of disc florets with large, strap-like petals. Despite an extremely large proboscis, butterflies tend to prefer small tubular flowers, such as those found on composites, *Buddleia* and *Lantana*. The proboscis is articulated from the base and also has a flexible 'knee' about halfway along its length which is clearly visible in this specimen. This arrangement gives the proboscis two 'degrees of freedom' of movement, rather like that seen in an Angle poise lamp. The proboscis walls are hollow like the cavity-wall of a house and can be filled with blood pumped in from the body. This allows the proboscis to be extended under pressure, but it relaxes under its own elasticity. Capsid bugs, such as that shown feeding at a flower of crown daisy *Chrysanthemum coronarium* (**51**), also have an articulated proboscis. Close examination shows, however, that the mechanism is completely different from the butterfly's. The mouthparts of the capsid bug, like those of a mosquito, consist of a pair of stylets surrounded by a jacket of more flexible material. It is only the latter that is retracted during feeding, allowing the fine needle-like stylets to be thrust into the soft plant tissues.

52. Many flowers of the mint family Labiatae are highly specialized for pollination by bees but there are exceptions such as the mint *Mentha longifolia* illustrated here. Its relatively simple, shallow flowers are accessible to flies and nectar-sipping beetles such as the soldier beetle *Cantharis*.

52

53, 54. Lilac flowers of the heath *Erica vagans* growing in damp oak-wood in the Cantabrian mountains of northern Spain (**53**). Ericas and heather *Calluna vulgaris* are important nectar sources in the upland regions of Britain and some parts of Europe, particularly in late summer and autumn. Although heather is said to be pollinated by the tiny 'thunder-fly' *Thrips*, its chief pollinators, like those of heaths, are honey-bees and bumble-bees. *Erica* pollen is shaken out of the anthers like salt through cylindrical pores which can clearly be seen in **54**. This photograph was taken in July when the bumble-bee breeding season had probably finished and it is noticeable that the *Bombus lucorum* shown is collecting nectar, but does not appear to be interested in pollen which is the normal source of protein for reproduction in bees.

55

55–60. Flowers of the borage family Boraginaceae are normally deep-tubular and are specialized for pollination by bees although long-tongued flies can also gain access to them. Heliotropes and forget-me-not *Myosotis alpestris* (**55**) have relatively shallow flowers that appeal to the short-tongued solitary bee *Andrena* shown in this photograph. The contracted throat of the flower of *Pentaglottis sempervirens* is just wide enough to admit the tongue of a bumble-bee (**56** on page 62) but keeps out other potential nectar thieves such as beetles. The vividly coloured alkanet *Anchusa strigosa* is here shown growing at the entrance to the gorge leading to the ancient city of Petra in Jordan (**57** on page 62). This flower has a deep, upright corolla tube which is favoured by long-tongued solitary bees such as the *Anthophora freimuthi* shown feeding from a hovering position (**58** on page 62). In contrast, the deep tubular flowers of the honeywort *Cerinthe major* (**59** on page 63) and comfrey *Symphytum officinale* (**60** on page 63) are inverted, with the mouth facing downwards so that only sufficiently agile and knowledgeable insects can gain entry, i.e. bees. These upside-down flowers have just enough of a lip around the entrance to the bell to provide a foothold for the clawed feet of the bee, in this case a *Bombus pascuorum* (**60**).

61

59

60

61

61. A spike of *Acanthus* flowers furnishes an example of a complex grouping of units that open in chronological order, from the base upwards. This arrangement is a reflection of the general law that the youngest structures in plants, be they leaves, buds or flowers, are usually at the apex. Other common examples of vertical inflorescences are willow-herb, foxglove, delphinium, hollyhock and bluebell. They are also found on trees such as horse-chestnut and sycamore, although in the latter case the spike hangs upside down. Bees visiting vertical inflorescences take account of their chronology by working from the bottom upwards, often moving spirally around the spike as they ascend. Towards the top, the younger flowers yield less and less nectar but it is more concentrated. Consequently the bee receives roughly the same energy reward as it goes and is less tempted to neglect the upper, smaller flowers. This is in the plant's interests because the youngest flowers

will normally be in the male phase, so that as the bee progresses from one spike to the next it takes pollen from male flowers to the receptive stigmas of female flowers.

62, 63. A flowering almond tree *Prunus dulcis* (**62**) exudes its own special scent which is rivalled, to the human nose at least, only by that of springtime orange blossom. But almond blossom is much more popular with bees. At any one moment a single tree may be visited by hundreds of individuals. The buzzing of these creatures as they go about their work only heightens the pleasure for anyone contemplating the scene, but from the tree's point of view there is a potential problem. A single tree has so many blossoms, and is so popular, that a bee will be tempted to stoke up with nectar and pollen from this single source without bothering to visit another tree. This behaviour may be economical for the bee but it reduces the likelihood of out-pollination for the tree. A tree laden with blossom presents to a

forager an unbroken mosaic of opportunities. Like a blackberry-picker who cannot believe his luck at being confronted by so many sumptuously laden bushes, it might be thought that the bee would hurl itself eagerly at the blossom in all directions. In practice however, the bee has to contend with other competing bees and it pays to work out a foraging strategy. The bee shown in **63** is surrounded by four flowers in addition to the one with which it is busy already. Its next move will be outward to one of these four. From this flower it will then turn randomly either right or left to a third flower and so on, so that on balance it will pursue a zigzag line across the mosaic. Every now and then it will perform a complete about turn and begin a new traverse across the canopy. This is the theory. In practice, with so many bees simultaneously engaged in foraging, it is quite usual to see bees ignoring a large percentage of flowers that lie in their path because they have been visited already and their nectar is spent.

62

64

64–66. Long-tongued flies rival bees in the frequency with which they visit flowers although they are usually less effective pollinators. Hoverflies are obligatory flower feeders. The drone fly (**64**) *Eristalis* has a 7 mm long proboscis but mainly visits shallow flowers like composites, ivy and blackberry. The proboscis of *Rhingia* is almost 1 cm in length allowing the insect to probe deep-throated flowers like white campion *Silene alba* (**65**) as well as shallower flowers like blackthorn *Prunus spinosa* (**66**). The *Silene* flower shown is a male but the

anthers poking through the mouth of the corolla have not yet dehisced, so the fly has not picked up any pollen. On the other hand, the *Rhingia* of **66** has a clear band of pollen running from the back of its head down the middle of its thorax. This is the zone that it is most difficult for the fly to groom with its legs. Flies are generally very scrupulous groomers but, unlike bees, they are interested in getting rid of the pollen, rather than retaining it on their bodies in special baskets. This is one of the reasons why flies make poorer pollinators than bees.

67–71. Amongst true flies or Diptera the most specialized nectar-feeding families are Nemestrinidae, an example of which is shown feeding on sea-lavender (**67**), and the bee-flies or Bombyliidae. *Bombylius major* looks like a tiny, furry mouse and is commonly found in springtime hovering around garden flowers such as *Aubrieta* and wild flowers like ground ivy *Glechoma hederacea* (**68**). From this hovering position it maintains a delicate toe-hold on the lower lip of the flower and is able to direct its proboscis above the protective line of hairs lying at the entrance to the corolla tube (**69**).

These hairs are designed to keep out smaller insects and possibly also to cut down the rate of evaporation of nectar from inside the flower. The tongue can be guided along the line made by the two white coloured anthers which are concealed beneath the upper lip but shine like beacons when viewed from below by an approaching insect. Members of another long-tongued family, the Empidae, are regular visitors to deep flowers. One was seen earlier in **34** (page 41), burying its proboscis into the 'nectar-tunnel' of a *Convolvulus* flower. Another, shown here (**70** on page 70), has not been deterred by the fringe of tentacle-like

hairs concealing the entrance to an alkanet flower *Anchusa strigosa*. This flower is normally pollinated by long-tongued bees such as *Anthophora*, shown earlier (**58** on page 62), but the tongues of large empids are almost as long. The body of the empid is covered with 'foreign' pollen since alkanet pollen is coloured white not yellow. A much smaller species of empid is probing the corolla of a forget-me-not *Myosotis* (**71** on page 71) which is normally pollinated by solitary bees like *Andrena* (**55** on page 61).

67

68

69

72–74. Typical 'fly' flowers, unlike those shown in **43–71** on pages 52–71, are shallow and relatively drab-looking. They are often green coloured and include dog's mercury *Mercurialis perennis* (which is visited mainly by small hoverflies), ivy and some tree blossoms like field maple *Acer campestris*. The pendulous flower spikes of maple are often clustered in springtime with the recently emerged St Mark's fly *Bibio marci* (**72**). This is a short-tongued species usually seen alongside similarly short-tongued craneflies, feeding on cow parsley and hogweed. Some flowers of the milkweed family Asclepiadaceae such

as the silk-vines *Periploca laevigata* and *P. aphylla* (**73** and **74** respectively) are especially adapted to pollination by short-tongued flies. These 'carrion' flowers are often malodorous and attract calliphorid flies which are also tempted to lay their eggs on them. Although it is the smell of rotting flesh that draws the flies to the flowers, they receive nectar by way of reward. In the case of the *P. laevigata* shown in **73**, the nectar appears to have dribbled out from the centre of the flower, through the slots at the base of each petal, and is now lying in tiny pools on the surface of the petals. One can imagine how the tongue of

the fly would easily be led in the reverse direction through the slots towards the central circle of stamens fused around the stigma. Between each pair of adjacent anthers is located a sticky protruberance, coloured purple in *P. aphylla* (**74**), on to which pollen is shed. This adheres to the proboscis of a visiting fly and may be deposited on to a flower of the same species that is visited subsequently. The mechanism is similar to that used by orchids which also attach a sticky pollen mass to the heads of their pollinators.

75

75, 76. Flowers of the pea family Papilionaceae are amongst the most specialized for pollination by bees. This is most evident in their shape. Unlike the majority of plant families considered so far in this chapter which are radially symmetric, i.e., have a generally circular shape, pea flowers are bilaterally symmetrical, i.e. two-sided. The nectar is concealed within a rather complicated corolla which can only be opened by bees possessing the necessary strength and behavioural instincts. The stamens are also concealed and only spring into action when the flower is triggered, dusting the visitor with pollen. Photograph **75** shows a honey-bee in the act of pollinating a broom flower *Cytisus praecox*. The bee is pushing with its head against the upper petal or keel and has used this purchase to provide leverage against the two wing petals with its forelegs. The stigma and stamens, previously hidden within the boat-shaped keel formed from the two lowermost petals, have sprung out against the underside of the bee's body. The bee is now 'riding' on the stamens, and using a series of rapid paddling movements of its legs to transfer the pollen first to the rows of hairs visible on the inside of the left hind leg, and from there on to the outside of the opposite leg where the growing blob of pollen is now seen. While all this is going on the bee is also taking its fill of nectar from deep inside the corolla.

77, 78. In some species of broom the stigma and styles are long and whip-like and the triggering mechanism is explosive. As a result the stigma and styles are flung over the insect's body and make contact with the safe sites on the back and head. The coiled style and stamens of a *Cytisus scoparius* flower that has just been triggered can be seen in **76** on page 75. In **77** a honey-bee has just triggered the mechanism and has almost become ensnared by the elastically coiled style. The stamens

with the longer filaments have thrashed against the bee's back, dusting pollen down the middle. A series of shorter stamens have come into contact with the underside of the body. The recoil of the style is powerful enough to catch an incautious bee in a temporary head-lock (**78**). Such involuntary detention does the flower no harm at all since it increases the contact time between it and its pollinator. Even after a flower has been triggered, honey-bees continue to visit it to

nibble at the anthers. Characteristically they hover from one flower to the next giving each a quick inspection, and ignoring any in which the petals have turned pale and begun to droop (as in **9** on page 19).

79. Honey-bees usually seem inept when pollinating *Cytisus*, and successful triggering of the floral mechanism such as that shown in **77** seems to rely on good luck rather than good management. In contrast, the bumble-bee *Bombus pascuorum*, as

in so many instances, is unerring in its dealing with the flower. Notice how, in **79**, the bee has used its greater strength and technique to force the wing petals right down with its middle and hind legs so that the tiny valvular processes at the bottom of the petals have been drawn out of the corolla tube. As in **77**, the upper set of longer stamens plus style have been whipped over the bee's back whilst the shorter set have pressed up against its under-body.

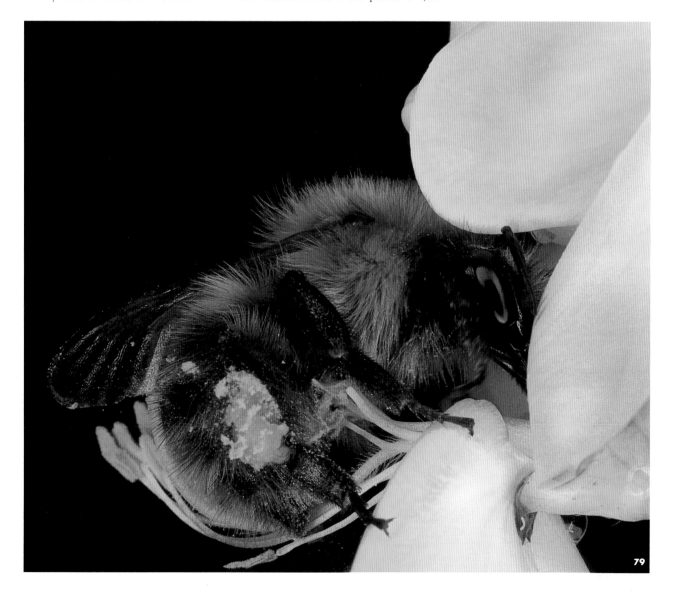

79

80. A bumble-bee *Bombus pascuorum* is shown pollinating a Spanish broom *Genista hispanica* which also operates an explosive mechanism, although in this case, as in *Cytisus praecox*, the stamens are shorter and only come into contact with the underside of the body. The flower lying adjacent to the one with the bee has already been sprung and will probably now be ignored.

81. This photograph shows a bladder senna *Colutea arborescens* being pollinated by a leaf-cutter bee *Megachile*. In this case, unlike the broom flowers featured in **75** and **76** on pages 74 and 75, the keel petals are fused except for a hole near their pointed ends. Depression of the wings and keel by the bee's hind legs produces a pumping action that forces the tip of the stigma out through the hole and on to the abdominal pollen basket of the bee, as can be seen clearly in the photograph. The stigma bears a stiff brush of hairs on to which the pollen of its own stamens has been deposited during the sliding action and this can now be transferred to the bee's body. At the same time, some of the pollen from another flower may be taken up by the stigma and, if it is receptive, the process of fertilization may begin.

82, 83. *Daboecia cantabrica* is a very large ericaceous flower shaped like an inverted bell. As in the totally unrelated comfrey and honeywort flowers shown in **59** and **60** on page 63, *Daboecia* is pollinated by bees which must suspend themselves from the bell by using its narrow upturned lip as a foothold. In **82** the nectar is being obtained legitimately from the flowers by a brown bumble-bee, probably *Bombus pascuorum*. Notice, however, the V-shaped incision near the base of the bell. This is a tell-tale sign that a nectar robber has already been at work. These are usually bees with slightly shorter tongues, which by-pass the normal route of entry into the flower and bite straight into the nectaries. A robber is seen at work in **83**, inserting its tongue into one of the holes made by either its own or, more likely, a previous culprit's mandibles. As likely as not, the robber will be surprised to find its tongue being tickled from within, since *Daboeica* also plays host to tiny St Mark's flies (**84**) which cluster eagerly around its nectaries.

85, 86. The foxglove is a perfect
example of a bee-flower: purple in
colour, with a very deep tubular
corolla that hangs upside down and
must be entered from below. The
mottling on the lip of the flower
serves as a nectar guide whilst the
recurved hairs repel smaller insects
and aid the inward passage of the
bee. The bumble-bee (**85**) is entering the
flower by the normal route. Typical
signs of illegitimate entry are shown
in the flowers in **86**, the narrow necks
of which are scarred by discoloured
incisions where nectar robbers have
been at work.

86

87–9. The long spur of the toadflax *Linaria triornithophora* contains nectar which is normally accessed through the mouth of the flower, after the upper and lower lips have been prised apart by a bee with the requisite strength and instincts. The bright orange throat boss serves as a visual guide to the position of the mouth. Try opening the mouth of a common garden snapdragon, which is a close relative of *Linaria triornithophora*, with your fingers and thumb and you will be surprised at the leverage that is needed. This gives an idea of the strength exerted by the bee as it wedges itself between the upper and lower lips of the flower (**88**). The incisions near the points of the spurs of the specimens shown in **87** betray the fact that these flowers have recently been subjected to the attentions of nectar-robbing bees. One of the latter is caught *in flagrante* in **89**. Invariably the approach of these nectar robbers is the same: they land on the flower, swivel their bodies upside down, scuttle down the spur, insert their probosces through the 'key-hole', and sip at their leisure.

87

89

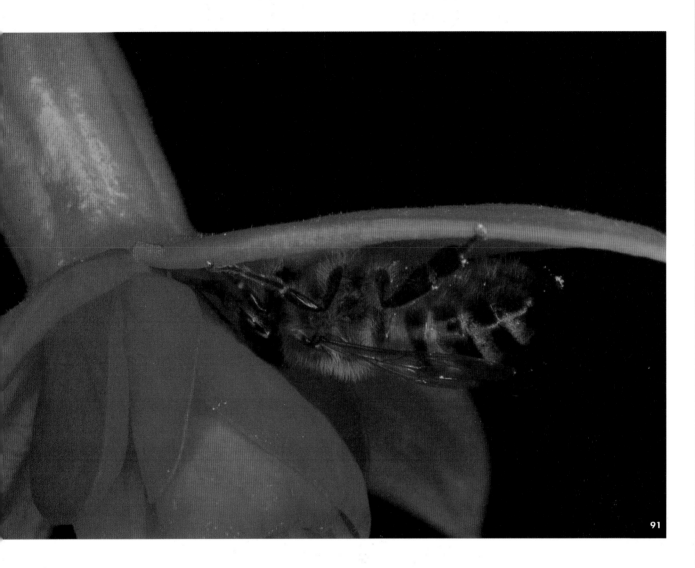

91

90, 91. The nectaries of a *Fuchsia* flower are deeply located at the bottom of the corolla tube as can be seen in **16** on page 24. Fuchsias in their natural environment are pollinated by birds but cultivated varieties in temperate countries are liable to become the objects of attention by bees. The honey-bee (**90**) is endeavouring to reach the nectaries via the normal route, and only the tip of its abdomen is left showing. Honey-bees however are resourceful insects and it is not unusual to find them losing patience with the usual route and attempting a short-cut by sliding their tongues between the bases of the outer row of petals (**91**). A glance again at **16** confirms that this cuts down the distance by at least 50 per cent.

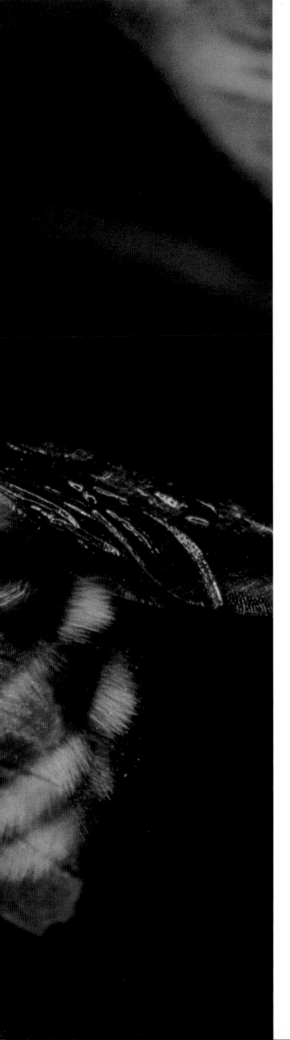

THE
VISUAL WORLD
OF AN
INSECT

I T WAS THE German scientist Sprengel who, 200 years ago, first recognized that floral colours are meant for their pollinators, and not for the human eye. Approximately a century later another celebrated biologist, the Swiss monk Mendel, formulated the elementary laws of genetics and laid the foundation whereby, if he chose, man could selectively breed almost any floral colour that took his fancy using little more than a painter's brush to transfer pollen between flowers. This distortion of normal evolutionary processes has produced objects that are beautiful but sterile and insects tend to avoid them, preferring the base species which they themselves brought into being using their own pollen brushes. A highly selected hybrid tea rose is nowhere near as attractive to a bee as its untidy hedgerow equivalent, the dogrose.

ABOVE *The cross shape of the flower guides the bee's tongue to the centre.*

OPPOSITE *The veins guide the bee's tongue into the flower.*

FLOWERS AND INSECT SENSES

Studies have now established beyond doubt that wild flowers make their appeal to insects primarily through their eyes rather than their sense of smell or touch. The attraction is so powerful that hungry insects will still be drawn to flower heads that have been taken out of their normal environment and placed in a completely artificial setting. Bees will pay just as much attention to flowers that have dropped to the ground, provided they are still fresh, as those that are left on the plant. Vision is not the exclusive sense; the fact that so many blossoms are heavily laden with fragrance indicates that chemical senses must also play their part. Night-pollinated plants such as stock *Matthiola*, wallflower *Cheiranthus* and honeysuckle *Lonicera* are particularly sweet-scented and this compensates for the ineffectiveness of visual cues in subdued light or darkness. Nocturnal pollinators are usually moths, but even bees, which are almost totally diurnal, use scent as a close-range cue to specific flowers. The smell of hawthorn *Crataegus*, white clover *Trifolium*, jasmine and *Cyclamen* seem to be especially potent. Bees in the hive can learn a flower scent from an incoming worker's body or even from the nectar that she has collected. The olfactory organs are located on the antennae and these are often lowered over the face towards the flower when feeding, as can be seen in photographs **17** (page 24) and **48** (page 57).

Although scent provides a useful close-range cue, it is easily understood why it could never rival vision as the primary method for detecting flowers. Fragrance strength fades rapidly with distance from the flower but, more importantly, the vagaries of the wind destroy its reliability as an indicator of distance or direction. We instinctively place our noses very close to a flower to appreciate its fragrance and the insect's sense of smell is probably no better than ours. This in itself supports the idea that it is the sight of a flower that first attracts the insect to it and only light can supply the exact information about three-dimensional shape, size, but above all colour, that the insect needs to recognize its quarry. Smell is too diffuse and it is only two-dimensional.

Although insects are highly sensitive to sound, and this form of signalling is frequently used as a means of communication, the plant world has never been able to invent a flower that rings like a bell. This notion is not entirely frivolous. There is one other form of energy in addition to light that can relay information about three-dimensional shape, size and distance and this is not sound, but ultrasound. Bats have been using sonar for millions of years as a means of reconstructing the sensual world around them. Some of the moths that the bat preys on have responded by developing ultrasonic discharges of their own in an attempt to confuse their persecutors, but sonar never seems to have been considered as a serious alternative to vision by foraging insects. However it has not been ignored altogether. Some bees, particularly those that are in the habit of hovering in front of flowers, such as species of *Anthophora* and *Anthidium*, emit ultrasonic pulses which vibrate the anthers causing them to release pollen like salt from a salt-cellar. An increasing number of plants are now known to rely on 'buzz-pollination' of this kind and in an effort to increase pollination success commercial growers have begun to mimic the bee's effect by beaming ultrasound into their greenhouses.

Flowers, or at least wild flowers, stand as a living testimony to the visual capabilities of the insect world but, when dealing with an organism which is so very different from ourselves, we must be careful in interpreting the evidence of our own eyes. A particular quality in a flower that might seem significant to us, might, however, bear no meaning for the insect. Take, for example, petal number. This is usually five, but four- and six-petalled flowers are also common. The constancy of this number throughout particular families almost amounts to a fixed law but it is not easy to think of any reason why these numbers in themselves should influence the insect's choice of flowers. On the other hand the positional relationships between the parts of a flower, such as the location of the stigma *vis à vis* the stamens, is of obvious importance because it affects the way the insect brings about pollination.

THE DESIGN OF THE INSECT EYE

Rather than relying too heavily on analogy with our own visual perceptions, it would be better to try to gain a better understanding of the way the insect sees the world by examining the mechanism of its eye. First of all, it is helpful to define exactly what is meant by an eye. It is possible to design an eye that perceives light but stops short at forming an image. Partially sighted people whose vision has become clouded by cataract, for instance, see the world in this way, and such a simple eye also serves the needs of caterpillars and many spiders. An eye of this kind is basically a transparent window in the skin which allows light to fall on to a collection of photosensitive cells. It can perform a variety of tasks including measuring light intensity, thereby informing its owner whether it is day or night, or whether the immediate environment consists of shade or open ground. It can also monitor crude differences of light across the window, such as those caused by the shadow of a moving object. If the object is moving rapidly it would be in the interests of the owner of the eye to take avoiding action, just in case the object were threatening. If a vague shadow loomed gradually larger in the window, this could mean that an object is approaching.

So we see that a considerable amount of information about size, movement and even shape can be gained from a simple lensless eye. The reader can obtain some idea of the range of possibilities by observing the to-ings and fro-ings of people in a room as seen through an interior door fitted with frosted glass. A lensless eye, even though it is primitive in design, can provide enough information to enable its owner to make important decisions relating to its welfare and self-preservation. The primal responses to light intensity and object movement continue to govern behaviour even when we come to consider a more sophisticated eye, capable of resolving an image.

It is usual to compare the workings of the human eye with those of a camera. Yet this does not diminish the remarkable convergence of design that these two instruments display. Both have a lens capable of forming an image on a light sensitive surface, the retina or the photographic film as the case may be. In both cases the amount of light entering the lens is controlled by an iris diaphragm which can be stopped down to prevent light flooding in, or opened up to allow in more light in dim-lit situations. A camera, however, can also bring objects closer or take them further away, depending on the focal length of the lens selected, and this is beyond the capabilities of the human eye. One could speculate on how human beings could cope with the distortions of perspective that would result if the eye worked like a telescope or a wide-angle lens.

Throughout the whole of the animal kingdom there appear to be only two basic eye designs: the human/camera type, which is found across the vertebrate spectrum from fishes, frogs and lampreys, through to snakes, birds and mammals; and the insect-type, found also in crustaceans and spiders. These two designs are so different that they must have been evolved completely independently. An insect's compound eye, as the name suggests, has a multitude of lenses studded across its curved exterior giving it the characteristic faceted appearance that we see, for instance, in photograph **94** on page 95. Depending on species, each lens measures only 0.02 mm to 0.1 mm in diameter. Like the cornea of the human eye it is made of transparent skin, and whenever the insect moults it casts off its old set of corneal lenses as well. Snakes and lizards also shed the outer coat of the eye during moult. Some insects also have a second layer of lenses located just beneath the first but these are not lost during moulting. Each double lens, or single lens as the case may be, guides a cone of light from the outside world down on to a group of pencil-shaped cells tied together like a bundle of sticks. These are the photoreceptors that absorb the light and convert its energy into an electrical signal which is conveyed to the brain. To absorb the light they are provided with a chemical pigment, and different photoreceptors within each bundle may contain different pigments. These differences provide the basis for colour vision, a subject that will be discussed at greater length in the next chapter.

IMAGE MOSAICS AND SENSITIVITY TO MOVEMENT

From the description given so far anyone could be forgiven for thinking that a compound eye containing ten thousand lenses, or lens pairs, should be capable of forming ten thousand separate images all over its surface. But such an eye would be completely unworkable. From time to time human beings inadvertently acquire double vision, although they are not usually in a position to give an exact description of the sensation. Suffice it to say that even two images are enough to make the world a very confusing place to be in. The insect avoids confusion on a potentially much greater scale by ensuring that each optical unit consisting of lens plus photoreceptor bundle, registers not an image but only a single spot of light. The brightness and colour of the spot are determined by the narrow cone of light rays coming into the unit from that particular part of the visual field that it samples. The sum total of all these light dots is an image mosaic whose graininess depends on the number of constituent elements in the eye.

A similar image mosaic is formed on the human retina but since the retinal cells can be numbered in hundreds of thousands, rather than hundreds *or* thousands, it is effectively grain-free. The human eye has evidently been evolved as an instrument for resolving the highest level of detail. The resolution of detail can be measured and expressed in terms of a quantity called visual acuity. The technique is quite simple. If the eye views a grid of equally spaced black and white lines from progressively increasing distances, a point is reached at which the eye is so far away and the lines of the grid are so close together that they can no longer be seen as separate images. The angle subtended to the eye between each pair of lines at this point of fusion is a measure of visual acuity. In the case of a human being with perfect sight, the angle is about one sixtieth of a degree. This is incredibly small and I sometimes wonder what it was in our evolution that necessitated an eye of such perfection.

Evidently, to judge by the relative paucity of retinal elements in their eyes, insects pay far less attention to detailed imagery than we do. Even the sharpest insect eyes can only boast a visual acuity of one or two degrees, poorer by a factor of a hundred than the human eye. If the human eye worked like an insect's, the embroiderer's art would be completely lost on it, to say the least. But there are more things between an insect's heaven and earth than highly refined vision. One of these is a remarkable sensitivity to movement and this is far more suited to the insect's real needs. Again, there is a convenient way of measuring how sensitive an eye is to movement. Imagine that you are invited to sit in a darkened room and given an image to view, let's say a picture in a book, with the aid of pulsed light from a stroboscope. If the frequency of the pulses is gradually turned up, eventually they fuse and the picture appears to be seen in continuous light. The point at which the intermittence of the light ceases is called the flicker-fusion frequency and for man it is about 50 per second. The flicker-fusion frequency tells us about the eye's ability to freeze moving images. It turns out that the corresponding figure in the case of insects can be as high as 300. So what registers as a movie to the human eye is seen as a slide-show by some insect eyes. Insects seem to be less interested in recognizing particular shapes or patterns than in seeing whether or not objects are moving, or getting bigger or smaller. The movement of the darkened edge of an object across the eye elicits more curiosity than the shape of the object itself. Depending on exactly how the edge is moving the insect can decide whether something potentially threatening is advancing towards it, or retreating away from it, or simply moving across its line of sight. The same sensitivity to movement can even be adapted to measuring distance. Grasshoppers deliberately scan the place to which they intend to jump by moving their bodies from side to side. This makes the image of the grass-stem, or whatever it is that has been fixated by the insect, move smoothly across its eye. By comparing the respective distances moved by image and eye, according to the principle of parallax, the grasshopper's brain can decide roughly how far away the landing site is, and how much effort will be needed to reach it by jumping.

92. The bulging eyes of insects produce virtually all-round vision, and considerable overlap of the visual fields in front lends the possibility of stereoscopic vision. The only part of the visual field not seen by these predatory robber-flies *Asilus crabroniformis* is the area obscured by their own body behind the eyes.

93

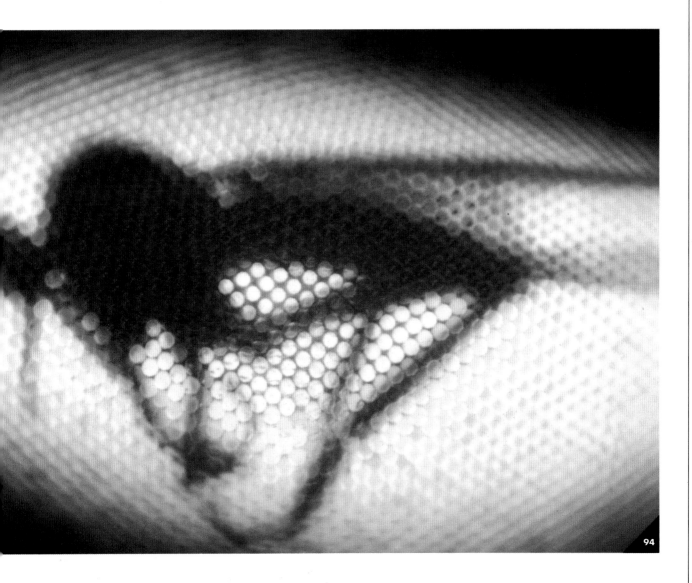

94

93, 94. The outer surface of the eye is covered by a fretwork of facets each of which is a minute corneal lens (**93**). The corneal lens focuses a narrow cone of light from a small angle of the visual field, down on to a series of photosensitive cells lying a short distance below the surface. Although insect eyes often look black, such as those of the flies in **92** on page 93, the pigment lies beneath the corneal lenses which are themselves totally transparent. A perfectly clear set of corneal lenses is sloughed off, along with the rest of the skin, during moulting. Photograph **93** shows an image of the front end of a stick insect on a leaf, seen through the corneal lenses. At higher magnification, the image of a robber-fly like the one in **92** is seen to be composed of a mosaic of light dots (**94**) each representing the average quality of the light entering a particular facet sampling a very small part of the visual field. These dots, like the pixels on a video screen, add up to form a highly granular image, so in general insect eyes are very poor at discriminating details of shape.

95

95–97. The narrow veins coursing down the petals of the flowers (**95** and **96**) help to guide the bees towards the mouth of the flower. But bees are remarkably short-sighted compared to human beings and the lines will only be visible by the time the insect has landed on the flower. The nectaries at the bottom of the imperial lily (**97**) shine like cat's eyes in the darkened interior of the flower. An enlarged view of these structures was shown earlier in **10** on page 20. Although they will help to guide any pollinator approaching from below, the eye-spots will only become clearly visible by the time the insect has reached the mouth of the flower.

95

96

98

98. If you walk out from the darkened interior of a building, such as this old abandoned church, into the sunlight and examine a clump of white flowers from which maximum light is being reflected, your eyes will experience a rise in light intensity of approximately one-thousand-fold. From being in a dark-adapted state, with a fully dilated pupil, the eye must quickly stop down the iris to the size of a pin-hole in order to protect the retina from the increased light. Insects also have a pupillary reflex but it takes much longer, a matter of minutes rather than seconds, for their eyes to adapt. In excess light, a screening pigment located at a deeper level within the eye migrates outwards to surround each photoreceptor, cutting down the light that it receives from the lens. Conversely, at night, the pigment in the eyes of nocturnal moths withdraws away from the photoreceptors, which become more reflective as a result and this is why moth eyes appear to 'shine' when viewed by torchlights in the dark.

99. The individual flowers in a spread of ox-eye daisies *Chrysanthemum leucanthemum* are clearly distinguishable from one another to the human eye from a distance of 5 to 10 m. The insect eye however is about a hundred times poorer than ours at discriminating neighbouring objects in space. Consequently a bee approaching such a stand of flowers will only be able to tell them apart when it is within 30 or 40 cm. In this case it is aided in its task by the yellow bull's-eyes of the flowers.

HOW INSECTS SEE FLOWERS

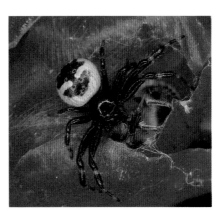

T HE PREVIOUS CHAPTER sketched out the workings of the insect eye but in this chapter we consider how these qualities, particularly the sensitivity of the eye to movement, contrast and colour, help insects to recognize flowers and to forage them more effectively. Imagine an insect making an approach to a flower. Its attention will have been drawn initially by colour, brightness and contrast and these qualities of light are all interrelated. On a dull day colour contrast is reduced so that brightly coloured flowers stand out less boldly from their surroundings than they would do on a bright day. Dull days are often also rainy, in which case foraging is out of the question anyway for most insects. Still air on windless days might seem to favour foraging but, given the choice, a bee would probably prefer bright, breezy conditions to still, dull ones. In bright sunshine, it takes a very stiff breeze indeed to deter a honey-bee or a bumble-bee: the insect simply hugs

ABOVE *An orange variety of the flowery spider* Synaema.

OPPOSITE *Tiny* Misumena *spiders trap insects visiting flowers.*

the ground more closely as it flies between flower patches, taking advantage of the 'boundary layer' of stiller air.

Anyone who has tried to take a close-up photograph of wild flowers will have some experience of the bee's dilemma. You find that the sun and the breeze play a game of cat-and-mouse. One moment the sun is shining and the flower is saturated with colour, but its petals are fluttering in the breeze and you cannot take the photograph. The next moment the sun disappears and you bite your lip in frustration but happily notice that the petals have at least stopped swaying. You now wait for the cloud to pass, prepare your finger on the shutter, but in the very instant that the light returns so does the breeze. Something rages inside you and very soon the flower becomes your sworn enemy.

RECOGNITION OF SHAPE AND DETAIL

The bee needs to be able to recognize the flower even when the breeze is blowing its petals out of shape. Shape, however, is less important than colour as a visual cue so this problem is probably not as acute as it might seem. In fact bees appear to be more strongly attracted to irregular than regular shapes, and the attraction is further heightened by movement. This is why bees often show more interest in flowers that are being stirred by wind than still ones. In the process the bee is simply exploiting those aspects of its vision to which it is best fitted: sensitivity to movement and colour. But at the same time it must be able to override its natural instinct to withdraw from any object that has a rapid, jerky motion. Far from being the slave of its own eye, the bee weighs up the evidence that it is receiving from its senses in the particular circumstances prevailing, and acts accordingly.

People often use the phrase 'making a bee-line' as a statement of positive intent, as though bees automatically 'made a line' to a particular flower head. This presumption is supported neither by the nature of its flight, which is zigzag, nor by the clarity of its vision, since the bee is hopelessly myopic. It is possible, at least theoretically, to calculate the distance at which an approaching bee will be able to recognize one flower head distinctly from another. This is the maximum distance from which it would make any sense to want to make a 'bee-line'. Before we can perform this calculation we need to know the visual acuity of the bee. This has been measured in an ingenious experiment in which bees were taught to associate a grid of parallel black and white lines with a reward, sugar. It is a credit to the determination of the bees that they were prepared to make a link in their minds between something as desirable as food and something as utterly alien as an optical grid. The experiment was designed so that the association would only be made if the lines were vertical. A grid of horizontal black and white lines was seen by the bees as unrewarding and therefore unattractive. If the bees spotted a vertical grid, they would fly towards it, expecting a reward. The experimental subjects were presented with a choice between vertical and horizontal grids, placed side by side but at increasing distances from the bees. At a certain distance when the bees were released, they could no longer execute a 'bee-line' to the preferred grid, because at that distance they could no longer distinguish a vertical grid from a horizontal grid.

From this experiment it was concluded that flying bees can only resolve with their eyes objects separated by at least 4 degrees in space. The images of objects placed closer together than this simply fuse into a blur. Applying these laboratory findings to field conditions might suggest that an approaching bee would only be able to see the images of two ox-eye daisies, growing side by side and each being 2 cm in diameter, when it was at most 30 cm away from them. The yellow 'bull's-eye' centre of the daisy obviously makes the task of recognition easier (see photograph **99** on page 99). For the bee, the 'far distance' of a spread of daisies, is anything greater than 30 cm! The bee is indeed very short-sighted. Once inside the 'near-field', the bee will be able to recognize the flower's shape, land on it, and then start resolving the shapes of individual florets on the disc. Again, allowing ourselves some licence, it should be possible to estimate the minimum sized floret that the visitor can discern. Assuming that the extended tongue is 1 cm long, and that when feeding the eyes are 1 cm away from the floret, the required answer is approximately 1 mm. In fact, few flowers have florets smaller than this, not only because it would be difficult to tell them apart, but also because it would be difficult to get an average sized tongue inside them.

Hoverflies seem to have the same aversion to 'bee-lines' as bees do. They are just as likely to approach a flower sideways-on as head-on. This may well be the more intelligent approach since it means that as the flower draws closer, its outline is more visible and moves more quickly across the nearside eye than the relatively 'blind' far-side eye. By comparing the image flow across the two eyes, the fly can then make estimates of distance and speed of approach. The situation may even be more complex than this. Sometimes hoverflies gradually alter the line of the body axis as they make their approach, as though they were scanning the flower with the nearside eye. As with the 'peering' grasshoppers considered in the last chapter, this scanning motion provides parallax information that could be used to refine the estimate of distance and approach speed. One is beginning to appreciate the complexity of the hoverfly's visual guidance equipment: it is like a theodolite and gyroscope rolled into one. Imagine a London trainee bus-driver exposed for the first time to the skid-pad. The instructor informs him that the object is not only to skid the bus to a halt, but more specifically to bring it to a halt at a bus-stop located on the other side of the skid-pad, and at which the examiner is waiting. One begins to wonder how hoverflies ever manage to land on flowers!

Colour vision

The question of colour vision has been reserved until last because it is in this area that our own experience conflicts most with that of the insect. The principles of colour vision are the same in all animals and to understand them it is helpful to think of the way in which a stage lighting engineer, using only red, green and blue lights, can reproduce virtually any colour required for the set on which he is working. Yellow

can be made from an equal mixture of red and green, plus a tweek of blue. Magenta results from equal mixtures of blue and red, and cyan from equal mixtures of blue and green. Adding the missing primary to any of these secondary colours restores white light. In other words the 'complementary' colour to magenta is green, and the 'complementary' to yellow is blue.

Now imagine the system turned inside out. The stage lighting engineer wishes to modify the lamps so that they can be used as a colour sensitive eye. Instead of driving mains electrical current into the lamp to make it beam out light, he changes the light bulb for a photosensor so that light beamed into the front of the lamp drives current out through the cable into an ammeter that measures it.

The filter on the blue lamp ensures that it will allow in only blue light, and so on for the red and green lamps. Instead of three lamps we now have three photoreceptors, each sensitive to a different part of the light spectrum. The instrument can now be pointed at any bright object and it will 'measure' the colour by comparing the currents generated by each of the photoreceptors. For example, if in a given experiment all three currents are the same, the object must be white because white light contains all the primary colours mixed together. If the red and blue receptors are silent, the colour must be green. If the green receptor is silent, and the red and blue receptors are equally active, the colour must be magenta. But if the green receptor is silent and the blue receptor is twice as active as the red, the colour of the object must accordingly be more bluish than reddish. Clearly this 'trichrome' system is capable of translating every conceivable nuance of colour into a simple, unambiguous ratio of three numbers. All we need is a 'brain' to compare the readings of the three ammeters.

In essence, this is the way human colour vision works. The retinal cells responsible for colour vision, the cones, contain either a red, or a green, or a blue photopigment. Bees also have trichrome vision, but the photopigments absorb at different wavelengths from their counterparts in the human eye. Bees are insensitive to red but they can see ultraviolet: their three photopigments absorb light more effectively at ultraviolet, blue or green wavelengths. Because the entire colour sensitivity spectrum of the bee is shifted downwards on the wavelength scale compared to ours, it is only rarely that human and bee colour perceptions will coincide. For instance, a poppy that shines vividly scarlet to us may well look black to a bee. Bees are in fact attracted to red flowers, but they are responding not to the red but to the ultraviolet light that is reflected from the petals at the same time. At the other extreme, insects can see ultraviolet markings on flowers, such as nectar guides, that are invisible to us. Bee-white is also quite different from human-white. White to a bee is an admixture of all the colours to which its eye is sensitive. This includes ultraviolet but leaves out the red wavelengths, so bee-white looks blue-green to human beings. Wearing blue-green tinted spectacles would give you some idea of what insects see through their compound eyes but this would only be an imperfect copy, since we cannot make our eyes sensitive to ultraviolet. Ironically, people who have had the lenses of their eyes removed surgically can see ultraviolet.

A bee sees many more 'blue' flowers than we do. Because it cannot see 'human' red colours such as pink, brown and cream which consist of mixtures of blue and red, they all register as 'bee-blue'. But the floral market does not only cater for bees. Some beetles, at least, can see red and bright red members of the Ranunculaceae such as *Adonis* and *Anemone coronaria* (see photograph **108** on page 110) are pollinated mainly by this group of insects. The eyes of butterflies have up to five different photoreceptors absorbing ultraviolet, violet, blue, green or red wavelengths. Flowers use the whole colour spectrum in order to attract pollinators, even though the market is dominated by solitary and social bees. Yellow, green and white flowers (to our eyes) are often favoured by flies, small bees and beetles.

Finally, a hypothetical question. What difference, if any, would it have made to the way in which flowers present themselves, if their pollinators had been totally colour-blind and only capable of seeing in black and white? From a simple mechanical point

of view it is not difficult to envisage such an eye. It would contain only one photoreceptor type containing a single photopigment that responded to light of all wavelengths. In essence this is the way in which the 'rods' in the dark-adapted human eye work. A monochromatic eye can register differences in brightness and contrast, but the great advantage of colour is that it can be used to produce contrast in condition of uniform illumination. Two objects, one yellow and one pink of equal luminosity, shape and size would be indistinguishable when photographed on black and white film. Therefore the answer to the hypothetical question is that colour contrast provides the flower with a tool for marking out a host of distinctions that would be invisible to insects if they had only monochromatic vision.

100–103. Flowers cover the whole colour spectrum and more, as far as insects are concerned, because insects can also see ultraviolet reflections from flowers that are invisible to human beings. The primary colours violet, green, yellow and red are here illustrated by *Viola* (**100**), *Thymelaea* (**101** on page 106), crown daisy *Chrysanthemum coronarium* (**102** on page 106) and crown anemone *Anemone coronaria* (**103** on page 107). Butterflies can probably see all of these colours, including ultraviolet, but bees are insensitive to red.

100

101

104–107. Cut-leaved lavender *Lavandula multifida* (104) is a typical 'bee' flower, in form and colour. Unlike most insects, bees can learn to associate colour with a reward, and they are much better at doing this task when dealing with colours in the ultraviolet/blue range of the spectrum. A pink flower such as the alpine *Dianthus monspessulanus* (105) looks blue to a bee because the bee cannot see the red component of the colour. Similarly the cream colour of these crucifers will probably be seen by the *Anthophora* bee as blue (106, 107).

106

107

108–110. Beetles can see red, and members of the buttercup family such as pheasant's eye *Adonis* and the crown anemone *Anemone coronaria* (**108**) are pollinated mainly by chafer beetles. Crown anemone is a typical 'pollen' flower, with a forest of stamens in which the beetle, in this case the specialist pollen feeder *Amphicoma*, scrabbles around collecting grains on its hairy legs and body. In this particular case the anthers have not yet dehisced releasing the pollen. Yellow flowers such as buttercups (**109**) tend to be visited by small bees, beetles, flies, ichneumon flies and sawflies. The yellow variety of the flower spider *Misumena* exactly matches the crown daisy on which it waits for prey and in this case camouflage has helped it to trap a drone fly (**110**).

111

111. The body of the carpenter bee
Xylocopa pubescens is violet-black to the
human eye, in other words it is almost
'colourless'. But our eyes fail to see the
reflected ultraviolet light that insects'
eyes see. Therefore to another bee, the
abdomen and wings of *Xylocopa* may
well seem as brightly coloured as the
yellow thorax does to us.

112, 113. The colour of tree blossom takes into account the fact that pollinators will usually be approaching from below and will see the flowers against a background of the sky. Open sky is dominated by short wavelength radiation, i.e. ultraviolet, violet and blue. Imagine that you were a bee capable of seeing ultraviolet, blue and yellow-green, but not red or orange. The colours that would show up best against the sky are the complementary colours, which in this case would be blue-green and green. In fact, most tree blossom is white: the blackthorn *Prunus spinosa* (**112**) is a typical example. Other familiar examples are hawthorn, elder, apple and horse-chestnut. Bees see 'human' white as green, which is indeed the best complement to the sky colours. One tree that significantly defies this rule is the Judas tree *Cercis* (**113**), but this is exceptional amongst European and Mediterranean plants.

115

116

114–116. One of the great advantages of colour vision is that it enables distinctions to be made between objects on the basis of light quality rather than light quantity. Two differently coloured objects standing side by side but reflecting light at equal intensities would be indistinguishable to a monochromatic eye but are instantly distinguishable to a colour sensitive eye. In the crocus (114) the attention of a pollinator is drawn to the reproductive parts, the stigmas and styles, by their yellow coloration which contrasts with the mauve of the petals. The crown anemone (115) has a central 'eye' of brighter colour against which the stamens are highlighted in order to attract the normal pollinator of this flower the chafer beetle *Amphicoma*. It is not surprising that relatively few green flowers exist because this colour contrasts least with vegetation. But there are exceptions. Spurge laurel *Daphne laureola* (116) is a favourite of honey-bees, but it probably gains a competitive edge by flowering very early in the season, even before spring has started, when few other sources of nectar are available.

117–123. Flower spiders do not spin webs to trap their prey but ambush it when it is foraging. Individual spiders show colour variations which adapt them to different flowers. Concealment coloration of this kind foils not only their prey but also potential predators such as birds. Photographs **117** and **118** show yellow and purple forms of a thomisid spider adapted to the yellow flowers of jujube *Zisyphus* and cardoon thistle *Cynara* respectively. The green colour of another spider (**119** on page 118) allows it to conceal itself amongst the leaves of the sticky rest-harrow *Ononis*. The flower spider *Misumena* has white, yellow and pink forms but successful predation is not entirely dependent on colour matching. These tiny white specimens (**120** and **121** on pages 119 and 120) are capable of catching prey much larger than themselves – a marbled white butterfly *Melanargia galathea* in one case – even though their colour matches neither the flowers nor the leaves. The spider *Synaema globosum* comes in at least two colour forms – with either red (**122** on page 120) or orange (**123** on page 121) patterning on the abdomen – but they do not necessarily match the flowers on which the spiders are hunting. The most common prey of *Synaema* are small species of bee. Since bees are red-blind it is unlikely that the colour banding of the spider is directed at them. More likely, the black and red pattern is a warning of distastefulness to potential bird predators.

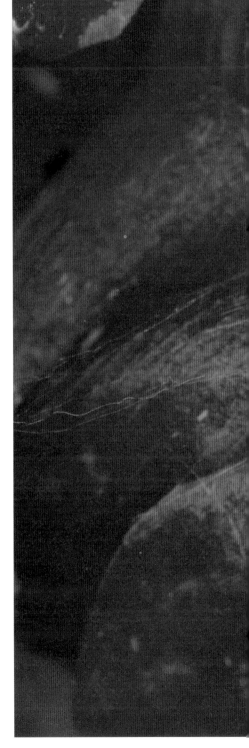

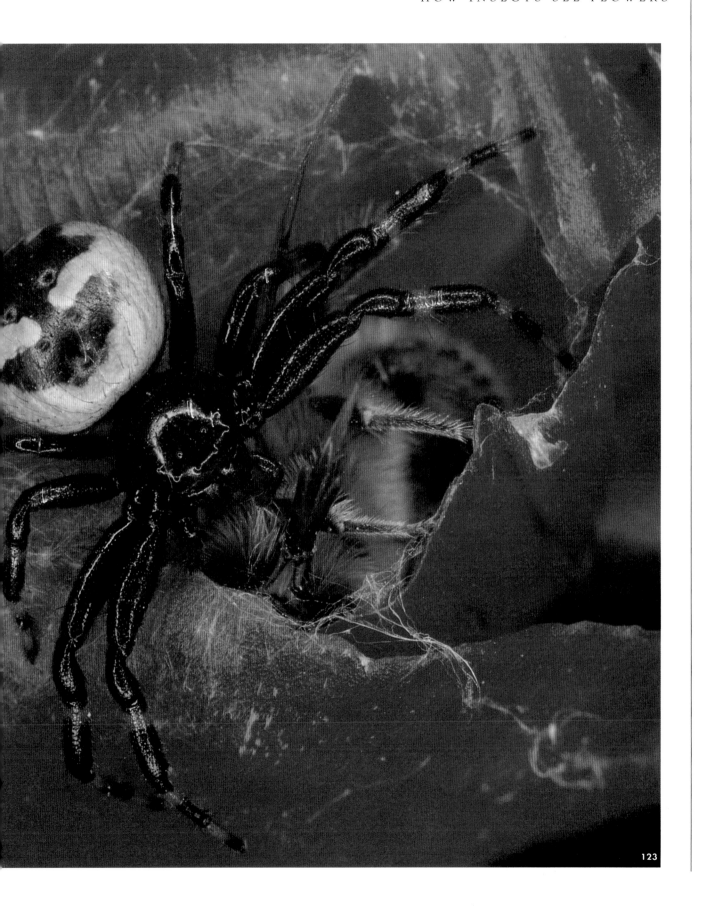

124, 125. To human eyes spiders' webs are only really noticeable in special conditions, for instance when they are covered in morning dew as in this hammock-web of a linyphiid spider (**124**). Some web-spinning spiders appear to take into account the peculiar sensitivity of the insect eye to ultraviolet light. The only natural sources of ultraviolet light are the sun and the sky and when insects fly up from the interior of vegetation they seek out an ultraviolet-bright light gap. The orb-web spinner *Argiope lobata* capitalizes on this by erecting its web across the gaps between vegetation. The *Argiope* web has built into it a zigzag spoke called a stabilimentum which scatters both ultraviolet and ordinary light when viewed obliquely against the sun (**125**). The fact that a spider's web is visible at all is mainly due to light scattering since the width of individual threads may be as little as 0.001 mm, which is well below the limits of even human visual detection. Although other functions have been ascribed to the stabilimentum such as camouflage, or acting as a warning signal to birds to avoid the web during flight between bushes, its main role appears to be that of a visual attractant to insects. This conclusion is supported by the finding that webs constructed without a stabilimentum are less successful in trapping prey.

125

6

INSECTS AND FLOWERS IN A DRY CLIMATE

NECTAR IS A vital secretion, not directly for the plant that produces it, but indirectly for its offspring. Energy and water are both invested in it. Whereas these are not usually a significant drain on reserves, in conditions of drought the water used for nectar production may have to be at the expense of the stem and the leaves. This can result in rather strange looking plants with shrivelled and dried-up stems but healthy flowers. As the season advances, with more and more plants having finished flowering, these resistant species take on an increasingly important role in furnishing the needs of local insect populations.

ABOVE An Amegilla *bee on a desert-flowering asclepiad.*

OPPOSITE Stizus, *a predatory wasp that feeds on nectar.*

INSECTS AND PLANTS OF THE DESERT

The 'desert bloom' is a familiar term that describes the springtime awakening of plant life in the desert. It coincides with the surge of activity taking place simultaneously at similar latitudes elsewhere but it is more dramatic because it is so short-lived and because of the transformation that it brings about in the otherwise barren landscape. Whether or not the bloom occurs is totally dependent on the preceding winter rainfalls, and in some years it is barely noticeable. Most of the springtime colour in the desert is due to annuals which germinate quickly when conditions are moist enough. They have no particular adaptation to dry conditions, and they must, as it were, 'make hay when the sun shines'. As the brief springtime gives way to summer, the evaporative power of the sun soon vitiates their meagre reserves and they shrivel. Seeds however, will already have been set and will remain viable throughout the summer and winter months, ready to germinate the following springtime.

Setting foot for the first time in the desert in March or April is a fascinating experience. You are taken aback by the profusion of pink campions, purple astragalus, yellow crucifers, white heliotropes and white woolly composites, most of them dwarf-like and springing straight out of the flat sandy earth. Dotted amongst these swards of annuals are grey-stemmed, switch-like perennials mostly belonging to the beetroot family Chenopodiaceae (photograph **126** on page 128). Their jointed, fleshy stems are salty to the taste, they have no recognizable leaves, and their flowers are minute but none the less attractive to bees. Unlike the annuals these perennial sea-blites and glassworts, many of which would feel just at home on marine mud-flats, are highly adapted to desert conditions.

A much greater variety of perennials can be found in the wadis or dry stream beds, in the fissures of rocky outcrops, or on the lower slopes of hills and mountain ranges. In short, anywhere that offers a little shade, and where the water-table lies within a few metres of the surface of the ground. Better drainage in springtime prevents salt-accumulation and instead of the salt-plants we see, for example, many members of the curious milkweed family, Asclepiadaceae. As their name implies, they secrete a concentrated milky latex, which bleeds freely from any tiny cut made in leaf or stem. This is such an effective means of retaining moisture against the evaporative force of the desert sun that one species of milkweed, *Calotropis procera* shown in photograph **131** on page 132 continues to flower in July and August with air temperatures soaring towards 50°C.

Another adaptation to water conservation is shown by various catchflies *Silene*, desert henbanes *Hyoscyamus* and rest-harrows *Ononis* which are extremely sticky to the touch because every part of the plant, bar the petals of the flowers, is covered in resin acting like a varnish against water loss. The same result can be achieved by investing the plant in an outgrowth of woolly hairs, a habit adopted by various dwarf species of cudweed *Filago desertorum*, plantain *Plantago ciliata* and sea-blite *Bassia eriophora*. The commonest adaptation seen in desert perennials is spininess. The plants take on a rounded, hedgehog-like appearance and often it is only possible to distinguish species by their flowers. *Zilla spinosa* is a mauve crucifer not unlike Lady's smock *Cardamine pratensis* in appearance and *Launea spinosa* resembles the common nipplewort *Lapsana communis*. It seems quite odd to see these familiar looking flowers sitting atop a tangle of grey, zigzag stems. Even more bizarrely, the desert mignonette *Ochradenus* looks more like an untidy, gigantic rush than the elegant spires made by the various species of *Reseda* as they grow by the roadsides in Western Europe.

Like the plants of the desert, desert insects show various degrees of adaptation to their environment. Many are extremely well camouflaged and difficult to spot, such as the sand-coloured mantis *Eremiaphila* that scuttles, beetle-like, across the stony ground, or the larva of the bagworm moth *Amicta* which lives in a case made of strips of vegetation and is almost invisible amongst the debris on the ground. Springtime brings an abundance of more familiar insects associated with the flowers, including species of solitary bee, bee-flies (Bombyliidae) and black-

and-red *Mylabris* beetles which, protected by their warning coloration, feed openly on the yellow composites. *Vespa orientalis*, the Middle-Eastern counterpart to the giant hornet *Vespa crabro* of Europe, is often heard hovering noisily around one of its favourite food plants, the above-mentioned *Ochradenus*. Equally noisy, and even larger, potter wasps (Eumenidae) such as the one shown in photograph **134** on page 133 feed on the milkweed *Gomphocarpus*.

SUB-DESERT AND DRY MONTANE ENVIRONMENTS

True deserts receive an annual rainfall of only 10–20 mm but they are only the extreme of a spectrum of environments characterized by low rainfall, high evaporation and very low relative humidity. Sub-desert conditions are encountered in most of the North African countries bordering the Mediterranean, and even in parts of southern Europe, notably Spain.

In these circumstances, one or two species of plant, depending on the locality, will often be found to be sufficiently resistant to the heat and dryness to continue flowering and producing nectar throughout the summer months. As a result they provide a natural focus for local insect populations. The common yellow ragwort *Senecio jacobaea* has gained a bad reputation as a rapidly spreading arable weed which is poisonous to grazing animals. Its tall, bushy inflorescences can easily be spotted in lowland pastures throughout Europe, standing alone amongst the livestock, everything else having been grazed to the ground. In the late summer months each clump of ragwort becomes an ecological niche. Bees, parasitic wasps, flies and the occasional butterfly visit it for its nectar and these in turn are fed on by various species of mantid including, in Spain, *Ameles spallanziana*, *Iris oratoria*, *Mantis religiosa* (photograph **162** on page 154) and *Empusa pennata*. Even in Britain, particularly during unusually dry summers, ragwort is one of the primary sources of nectar in the late season. Although there are no British mantids, the spider *Enoplognatha* shown in photograph **164** on page 155 often spins its web there.

In southern Spain and Morocco a similar role to

that of *Senecio jacobaea* is played by the tiny carline thistle *Carlina racemosa* (photograph **158** on page 152). Whole tracts of landscape that form natural grazing land in the springtime months gradually become transformed into a sea of pale yellow as the summer advances and the drought-resistant thistle takes hold. The flower heads are popular with bees but the rest of the plant is so brittle that it crackles underfoot. The main predators making use of this particular ecological niche are mantids and assassin bugs. A pair of the latter are seen in photograph **160** (page 153) contesting the carcass of a bee.

Thistles of various kinds are well suited to fill the 'nectar void' that tends to appear in late summer in Mediterranean countries when most other flowers have shrivelled and died. They are often late flowering, possess an innate resistance to dehydration and, on account of their spininess, are often resistant to grazing. In the sub-desert region between the southern Atlas mountains of Morocco and the edge of the Sahara, one of the few plants to continue flowering into August is the 'wire-netting' plant *Launea spinosa*, referred to earlier in this chapter. The same prickly composite can be seen growing in the almost waterless environment of the Canary Islands off the coast of North Africa. In the Atlas mountains themselves, the late summer nectar market consists mainly of species of parasitic hymenoptera including spider-hunting pompilids, sphecids and ichneumonids which scour the barren uplands in search of prey with which to provision their nests, but feed themselves on a much more benign diet of nectar. Wherever the spiny thistles *Echinops polyceras* and *Cynara* grow the local insects will be found clustered upon them (see photographs **146–157** on pages 142–151). The *Cynara* flower head depicted in photograph **146** seems totally repellent to mortal flesh, but I have it on the authority of local goatherds that, whereas sheep ignore it and goats will only nibble at its blue, tufted petals, a hungry donkey will chew off the whole head, barbs and all.

The northern slopes of the Atlas are rather greener than the southern because they receive rainfall from the Atlantic throughout the summer months in the form of flash-storms, and it is here that evergreen oak and cedar forests prosper. Even in the more extreme environments at altitudes of 2000 m or more it is still possible, in August, to find plants in flower such as dwarf campanulas, spiny *Centaurea* species, *Euphorbia* species and the aromatic herb *Ruta*. The latter grows in abundance in the rock-strewn clearings of the cedar forests and a clump of it is a good place to spot, amongst a whole variety of parasitic wasps, the gigantic cuckoo wasp *Scolia flavifrons* eagerly belly-flopping from one flower to the next.

The cedar forests of the High Atlas are a fitting place to end this tribute to insects and flowers. My wife Zena and I first began our pilgrimage to these bewitching mountains nearly a decade ago and almost every year since we have rattled our aged Land-Rover through Spain and by ferry across the Straits of Gibraltar to get there. In the evenings the air is still and filled with the sweet incense of the cedars. Perched on top of these gigantic trees, ravens call hoarsely to one another across the clearings. As night falls, the muffled churring of nightjars grows louder and louder. A shriek from within the forest tells us that a family of baboons is on the move. But the most stirring sound of all makes me reach instinctively for my flash-camera. I cannot hear the night-time pealing of the Italian cricket *Oecanthus pellucens* without seeing in my mind the image of its vibrating lace-like wings raised elegantly above its body and wishing to capture it on film. Zena knows the signs, from long experience. Pulling the cotton sheet around her, and preparing for bed, she calls down from the roof of the Land-Rover, "See you in the morning, dear".

126, 127. Wadi rum in the Jordanian desert (**126**). This desert, like the Negev desert of Israel to the west and the Iraqi and Arabian deserts to the east, receives an annual rainfall of only 10–20 mm, most of it coming in the winter months. The grey-green brush-like vegetation on the sandy floor consists of saltworts, succulent members of the goosefoot family Chenopodiaceae which are able to withstand the high salt content of the surface soil. In between, in springtime, is a host of annuals which flower briefly before the moisture is driven from the soil by the advancing sun. The jointed stems of saltworts continue to produce tiny flowers throughout the summer and are visited by bees primarily for their pollen (**127**).

127

128–130. Spring flowers grow in profusion in the rocky outcrops above the desert floor. There is shade for at least part of the day and better drainage prevents salts from accumulating to toxic levels. Typical annuals include mayweed *Anthemis* (**128**) and the poisonous golden henbane *Hyoscyamus aureus* (**129**) whereas the perennial caltrop *Peganum harmala* (**130**) can also withstand the more testing conditions on the desert floor.

134

133

131–134. The asclepiad *Calotropis procera* (**131**) can withstand extreme aridity. The whole plant is filled with a milky latex providing resistance to dehydration, and as a result flowering can continue virtually the whole year round. The corolla bears five spurred projections surrounding the central fused column of stamens (**132**) and it is into these that the nectar collects. These are the parts of the flower specifically targeted for example by ants (**132**) and the anthophorid bee *Amegilla* (**133**). Another asclepiad *Gomphocarpus* (**134**) has reflexed petals and the same ribbed corona as is found in *Calotropis*. The gigantic potter wasp *Delta dimidiatipenne* shown in this photograph is specifically probing a coronal projection for the nectar which it contains.

135–139. Like asclepiads, spurges (family Euphorbiaceae) also secrete a milky latex helping them to tolerate extreme aridity. The tree spurge *Euphorbia dendroides* (**135**) becomes dominant locally in dry rocky areas in Mediterranean Europe, North Africa and the Canary Islands. The spurge flower carries four half-moon shaped glands, on to the surface of which nectar is secreted, and it is from this that the anthophorid bee (**136**) and ant (**137**) are feeding. In late summer, spurges provide nectar for parasitic wasps such as the spider-catching pompilid *Hemipepsis* (**138** on page 136) and the sphecid *Sphex* (**139** on page 137). Note how the sphecid's thorax and legs are covered in fine, silvery hairs to reflect the desert sunshine.

140

140–142. Some desert insects also benefit from human cultivation in the oases. Oases usually become established along river beds. The thrifty use of precious water reserves allows year-round irrigation of the rich valley soils which are therefore able to support a vast range of root crops, fruit and maize for human consumption and blue alfalfa which is grown as animal fodder (**140**). Alfalfa is a rich source of nectar and access to it means that species such as the scarce swallowtail *Iphiclides podalirius* can survive in habitats well beyond their normal range (**141**). The alfalfa fields become the mainstay of a local ecology, and desert predators such as the mantid *Blepharopsis mendica* (**142**) become drawn in from the surrounding dry hillsides and wadis.

143. Desert and sub-desert species of yellow rest-harrow *Ononis* are covered in a sticky glandular secretion which acts like a varnish to cut down water loss. These may be found flowering in dry stream beds even during the hottest months of the year and are visited by bees for both nectar and pollen.

144, 145. The succulent habit is another adaptation shown by plants of arid environments. Familiar succulents found in salt-rich desert areas are *Aizoon* and *Mesembryanthemum* whilst on higher rocky ground in North Africa and the Canary Islands *Aeonium* (**144**) sends up huge, brilliant yellow spikes of flowers pollinated by bumble-bees and honey-bees (**145**).

145

146–149. The bright spiny heads of the thistle *Echinops polyceras* growing around Lake Tislit 2000 m up in the High Atlas mountains lend colour to an otherwise barren and lonely setting (**146**). From springtime onwards the enclosing mountainsides are parched and the lakeside flowers become increasingly important to local insects. Many of them are hymenopterans such as the bee-killer wasp *Philanthus triangulum* (**147**) and the large colourful pompilid *Hemipepsis mauritanica* (**148** on page 144), both of which are short-tongued. The digger wasp *Bembex rostrata* (**149** on page 145) has a beak-like extension of its mouthparts enabling it to probe much deeper into the florets. The abundance of these parasitic hymenoptera in this unusual habitat probably makes them the main pollinators of the lakeside plants.

146

149

151

150–157. The northern slopes of the Atlas mountains receive more regular rainfall from the Atlantic than the southern slopes and are lusher in appearance. Most of the summer rainfall, however, has a high run-off rate leading to flash-floods. Soaring summer temperatures also mean that evaporative rates are high. Consequently, the vegetation benefits little from any summer downpours that do occur and they are referred to locally as 'devil's rain'. In the stony clearings within the cedar forests of the High Atlas the viciously spined cardoon thistle *Cynara* (151), a late flowerer, is one of the main food sources for local bees like the giant carpenter *Xylocopa violacea*, the leaf-cutter *Megachile* (150) and the miner *Halictus* (152) along with the digger wasp *Stizus* (153 on page 148). The butterflies *Mesoacidalia aglaja* (dark green fritillary) and hermit *Chazara briseis* (154 on page 149) are also regular visitors. The bush-cricket *Ephippiger* sings from within the cage of spines formed by the flower heads (155 on page 150) which also provide concealment for the mantids *Ameles* (157 on page 151) and *Mantis religiosa* (156 on page 150).

152

153

154

155

156

158

158–160. A carpet of the dwarf carline thistle *Carlina racemosa* covering hillsides in northern Morocco (158). Despite the dry summer heat, the thistle continues to produce nectar and is much frequented by bees. Lying in wait for them are the mantid *Ameles* (159) and the assassin bug *Rhinocoris* (160).

159

161–164. The ragwort *Senecio jacobaea* growing in dry pasture in southern Spain (**161**). Like *Carlina racemosa* (which can be seen growing around the ragwort in this photograph) *Senecio* is an important source of nectar in late summer when most othert plants have ceased flowering. Because it contains poisonous alkaloids, it is highly resistant to grazing. In Spain, four species of mantid, including *Mantis religiosa* (**162**), use it as a lair to catch pollinating insects. Ephippigerid crickets are also well matched to it by their coloration (**163**) whilst in Britain the spider *Enoplognatha* (**164**) regularly spins its flimsy web amongst *Senecio* flowers. In this case the spider has caught an empid fly.

FURTHER READING

FAEGRI, K. AND VAN DER PIJL, L. (1978) *The Principles of Pollination Ecology*, Pergamon, Oxford/New York/Toronto/Sydney/Paris/Frankfurt.

FREE, J.B. AND BUTLER, C.G. (1959) *Bumble Bees*, Collins, London.
An old text, long out of print, but an excellent account of its subject.

O'TOOLE, C. AND RAW, A. (1991) *Bees of the World*, Blandford, London, Facts on File, New York.
A scholarly account, fully illustrated with line drawings and colour photographs.

PROCTOR, M. AND YEO, P. (1973) *The Pollination of Flowers*, Collins, London.
Along with Faegri and Van der Pijl, the classic texts on pollination biology.

The following are excellent technical reviews:

CRAIG, C.L. AND BERNARD, G.D. (1990) 'Insect attraction to ultraviolet-reflecting spider webs and web decorations', *Ecology*, Volume 71, pp. 616–623.

KEVAN, P.G. (1972) 'Floral colours in the high arctic with reference to insect–flower relations and pollination', *Canadian Journal of Botany*, Volume 50, pp. 2289–2316.

KEVAN, P.G. AND BAKER, H.G. (1983) 'Insects as flower visitors and pollinators', *Annual Review of Entomology*, Volume 28, pp. 407–453.

MENZEL, R. AND SCHMIDA, A. (1993) 'The ecology of flower colours and the natural colour vision of insect pollinators: the Israeli flora as a study case', *Biological Reviews*, Volume 68, pp. 81–120.

ACKNOWLEDGEMENTS

As always, I would like to thank my wife Zena for sharing my enthusiasm in this project, and for the infinite patience that she has shown during our various travels in search of material. She seems to have a gift for turning up at the very moment when I have begun to despair of ever finding my quarry and remarking nonchalantly, "Oh, there's one of those beetles you've been looking for." I am also grateful to Richard Wang for the literature searches that he has conducted on my behalf, and to Chris O'Toole of the Hope Entomological Museum, University of Oxford, and Sally Corbet, of the Zoology Department at Cambridge University, for help in identifying insects. Mrs Jane Seymour-Shove typed the manuscript and the Central Audio Visual Aids Unit of the University of Cambridge processed the photographic film with their usual promptness and efficiency. Finally, I am grateful to Stuart Booth, Consultant Editor at Cassell, for the faith that he has shown in the project and for his constant encouragement during the writing of the book.

INDEX

Numbers in **bold** refer to illustrations (by caption numbers)